D1548323

Benchmarking in the process industries

Benchmarking in the process industries

Munir Ahmad and Roger Benson

Published by
Institution of Chemical Engineers,
Davis Building,
165–189 Railway Terrace,
Rugby, Warwickshire CV21 3HQ, UK
IChemE is a Registered Charity

Reprinted 2000

ISBN 0 85295 411 5

Printed in the United Kingdom by The Cromwell Press, Trowbridge

Acknowledgements

The authors would like to acknowledge the contributions made by numerous colleagues through discussions, comments or meetings on the various topics included in the book. There are far too many people to acknowledge by name but we have attempted to mention a few as, without their help and support, the task of completing this book would have been impossible.

We undertook this project in addition to our normal workload and, as a result, families were usually neglected. Munir is very thankful to his wife, Maryam, and children, Aryan and Monica, who always wanted to do something but he was busy working on the book at his own or Roger's home. Roger is also grateful to his wife, Kathlyn, and children, Richard, Louise and Martyn, who patiently accepted early morning disturbances working on the book.

We would like to thank Professor Andy Matchett for the invitation to write a book on the subject for the Institution of Chemical Engineers. It has been a good experience for both of us to learn from each other during this project.

Roger is deeply indebted to his colleagues at ICI Manufacturing Technology who have helped in all stages from the book's foundation in 1995 through to the date of publication, and they will see their ideas within the text. In particular, he would like to acknowledge the creative contributions of Tony Conning and Brian Hull for much of the original thinking and Pamela Dinsdale for her patient typing. Other ICI colleagues who have made a significant contribution are the ICI Acrylics Manufacturing Excellence Team led by Dr Jeff Davis, and Sandy Anderson for his leadership. This is further added to by both the patience and knowledge provided by all the process plant managers inside and outside of ICI who have made a tremendous contribution to this approach of benchmarking.

During Roger's period at the DTI Innovation Unit, under the leadership of Dr Alistair Keddie, he was appointed one of the judges for Britain's Best Factory Award. This provided much of the early insight into benchmarking process plants. Roger would like to acknowledge the contributions made by other fellow judges Professor Colin New, Marek Szwejozewski, Malcolm

Wheatley and John Budgen. Further stimulation was provided by colleagues at Imperial College, University of Teesside and Newcastle.

We would like to thank the support given to us by our colleagues in the European Process Industries Competitiveness Centre (EPICC) and, in particular, Ian Smith who helped on the computer technology side. We would also like to thank colleagues in the Process Manufacturing and Design Section, School of Science and Technology, University of Teesside for their support and encouragement, and the postgraduate students (including those on the Masters course in process manufacturing management) for their contributions in literature searches and input through tutorial sessions.

Finally, we are grateful to Claire Aitken for her editorial input, Tracey Donaldson at the initial stages of the project and, most of all, to the Institution of Chemical Engineers for giving us this opportunity of putting our ideas on the subject in context so that students, teachers, trainers, consultants, industrialists and the process manufacturing industries as a whole, can benefit from this work.

Munir Ahmad and Roger Benson

Contents

Introduction

1

Process manufacturing covers all industries involving a chemical or physical change in the manufacturing process such as chemicals, pharmaceuticals, food, steel and mineral processing. These are recognized as being the most successful manufacturing industries in the UK where total sales in 1997 were £33 billion, with profits of £4.8 billion, and they are fast becoming the largest single export group[1,2]. The process industries are very familiar with the routine measurements of the financial, safety and environmental performance of its operating plants. It is the industry which fathered chemical engineering as a discipline and is at home with the measurement and benchmarking of chemical performance.

Surprisingly, it is much less familiar with measuring and benchmarking its own process manufacturing performance. The techniques for benchmarking manufacturing performance have primarily been developed and applied in other manufacturing industries such as electronics, retail and automotive. The argument often used as to why this unfamiliarity occurs is that the process industries are different and these benchmarking techniques do not apply.

It is our contention and experience that this is not true. Benchmarking is possible in the process industries by adding value that is essential for the future competitiveness of the industry. This book describes how to measure and benchmark the manufacturing performance of a process plant as a first step on the journey to performance and competitiveness improvement (see Figure 1.1).

Benchmarking is a structured process[3–6] comparing the performance of similar manufacturing assets against the best in the world, with the intention of learning and hence continuously improving. It is based on our experience of benchmarking over 200 process plants in many countries and different industries. The largest had over 600 employees while the smallest just two. The performance measures and approaches have been validated in a variety of international plants demonstrating how value is added.

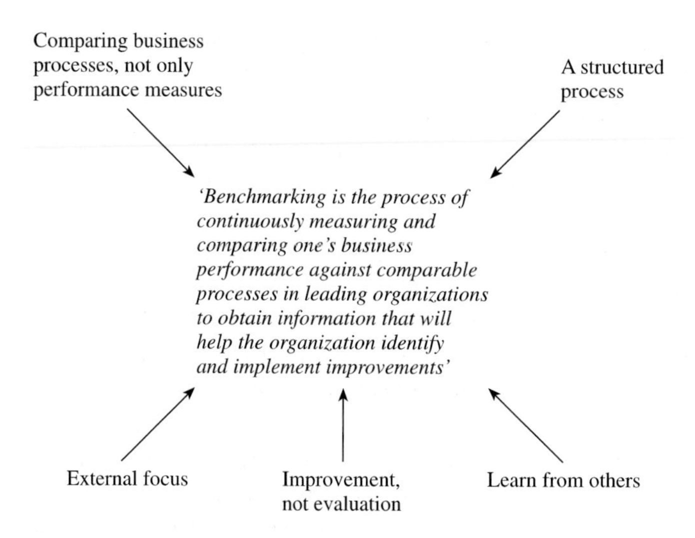

Figure 1.1 The definition of benchmarking (reproduced by permission of ICI Technology)

Furthermore, experience has shown time and again that the rate at which potential improvement in manufacturing is released is in excess of the rate of growth of the world market for most of the chemicals which are produced.

It is quite common in the process industries to determine during the first benchmarking exercise that:

- the hidden plant may well be in excess of 30% of output;
- stocks may be reduced by 50%;
- the fixed cost can be reduced by anything up to 20%;
- potential variable cost reduction by a further 10%;
- customer service is poor.

Given that they may often be achieved with minimal to zero capital expenditure, and are delivered through the people of the plant itself, they provide very interesting opportunities.

This implies that as the process industry moves to world-class manufacturing to complement its world-class process performance, the winners will succeed and profit and those who do not follow this plan will eventually close, as shown by trends in other industries — for example, Fortune 500 reveals that during 1960–1985 (25 years) 150 companies dropped out compared with 143 companies during 1985–1990 (five years)[7–10].

Chapter 2 describes the measurement framework to benchmark a process, which is based upon three elements:

- measuring the output of the process — that is, the performance;

- measuring the input practices that are applied to deliver these outputs;
- measuring the culture and softer issues that ensure that the practices are applied and the performance is delivered.

The actual choice is influenced by the drivers and priorities of the business together with the culture and values of the operation. Consequently, this book does not offer a prescribed template for benchmarking and scoring performance. While it provides the framework, measures and industry world-class targets, the actual detailed framework and scoring must be determined by the individual company. There is no one answer for all process industries.

Chapter 3 assumes the framework has been selected and focuses on the performance measures. These are the outputs/deliverables of the manufacturing performance. In this context, manufacturing is taken as all elements under the control of the operations manager. As such it may be from raw materials through to delivery of the final product to the customer, or the narrower scope of a production manager. It addresses the question of where to find benchmark data. Some benchmarks are shared as guidance for the less experienced and a means of scoring is suggested.

Chapter 4 broadens the measurement into the interpretation of the measures in financial terms. It introduces the concepts of the 'hidden plant' and the opportunities in variable, fixed and stock costs. In the process, it provides evidence to challenge some of the industry myths such as stock results in better customer service.

Chapter 5 expands on this area and indicates how the high-level benchmarks provide the route to deep-drill into the immediate priorities. Route maps are provided for each measure. While only indicative, these are the bridges between benchmarking and the more 'conventional' technical functions in a process company.

Chapter 6 focuses on benchmarking the practices. These are the practices and procedures that operating units adopt to improve performance. While several alternative models are briefly discussed, the common characteristics are that they may be scored and plotted against the performance score to fully position performance.

Chapter 7 discusses the soft/cultural aspects of an operation. Most people have experienced the 'buzz' when visiting a good operation. The characteristics that create this perception are very important when forming an overall view of process manufacturing performance. It is a sort of sanity check. This chapter provides a framework for measuring this characteristic.

Chapter 8 consists of a set of examples to help the reader understand how to interpret the mass of information generated in a benchmarking study and focus on the priorities to start the improvement process.

Arising out of the measurement of process manufacturing performance and efficiency is a growing realization of the need to increase the 'agility' of process manufacturing if the industry is to remain competitive. Chapter 9 focuses on this area and gives some pointers for the future based on the learning from the UK Foresight study.

Finally, Chapter 10 speculates on how process technology can develop to impact on the existing accepted world-class manufacturing performance targets. It attempts to define 'tomorrow's' world-class standards in the belief that tomorrow is not actually very far away in process industry terms.

In support, the Appendices provide a summary of the key measurement definitions and a specimen data collection form. This will need tailoring by the individual reader and a scoring system will need to be applied that reflects the individual business strategy.

The applications of the methodology and approach outlined in this book are very far reaching. For example, they may be applied to:

- existing assets to measure the improvement potential;
- proposed new assets at the design stage to ensure that the manufacturing performance that it is being designed for is at least as good as the best that is currently being achieved within the business or industry;
- as part of the diligence process in an acquisition, to determine the state and the competitiveness of the assets being purchased and hence to influence the price that is negotiated;
- to measure the performance of independent change consultants that may be employed by the company. Independent measurement of the manufacturing performance before and after the change programme is a very effective way of measuring delivery and thus payment.

The benchmarking approach[11–17] is all embracing and one that the process industries could and should apply more widely. It is our opinion that improving manufacturing performance is the next big opportunity for the world of chemical processes, particularly when the nature of process industry business is changing so rapidly.

References in Chapter 1

1. DTI, 1997, *Competitiveness Report — 1996–1997* (DTI, UK).
2. UK Government, 1994, *White Paper on Competitiveness* (HMSO, UK).
3. Baldridge, M., 1996, *Winners Profile Booklet* (Malcolm Baldridge National Quality Award Office, National Institute of Science and Technology, Gaithersburg, MD, USA).

4. Bendell, T. *et al.*, 1993, *Benchmarking for Competitive Advantage* (Pitman Publishing).
5. Camp, R.C., 1989, *Benchmarking — The Search for Industry Best Practices that Lead to Superior Performance* (ASQC Quality Press, Milwaukee, USA, ISBN 0 87389 058 2).
6. Cavanaugh, T., 1995, Quality strategies 95: Benchmarking goes global, *Chemical Marketing Reporter*, 247(15): SR16–SR17.
7. 1997, Fortune 1000 ranked within industries, *Fortune*, April 28, pp. 44–66.
8. Gardiner, K.M., 1998, Globalization, integration, fractal systems and dichotomies, *Proc of 8th Int Conf on Flexible Automation and Intelligent Manufacturing* (Begell House Inc, New York, ISBN 1 56700 118 1), pp. 15–27.
9. Deming, W.E., 1986, *Quality, Productivity and Competitive Position* (MIT Centre for Advanced Engineering Study, Cambridge, Massachusetts, USA).
10. Welch, J., 1993, Lessons for success, *Fortune*, Feb 25: 8.
11. Fitzgerald P., 1996, Benchmarking pays off, *Chemical Marketing Reporter*, 249(15): SR16–SR17, April 8.
12. Lema, N.M. and Price, A.D.F., 1995, Benchmarking performance improvement toward competitive advantage, *J of Management in Engineering*, pp. 28–37.
13. Zairi, M., 1996, *Effective Benchmarking — Learning from the Best* (Chapman and Hall, UK).
14. Zairi, M. and Leonard, P., 1994, *Practical Benchmarking — The Complete Guide* (Chapman and Hall, UK).
15. Voss, C.A, Chisa, V. and Coughlan, P., 1994, Developing and testing benchmarking and self-assessment frameworks in manufacturing, *Int J of Operations and Production Management*, 14(3): 83–100.
16. Watson, G.H., 1993, *Strategic Benchmarking — How to Rate Your Company's Performance Against the World's Best* (John Wiley and Sons Inc, UK).
17. Appleby, A. and Prabhu, V., 1998, Implementing best manufacturing practice: are we winning?, *3rd Int Conf Managing Innovative Manufacturing* (University of Nottingham, UK), pp. 71–78.

2 Measuring process manufacturing performance

2.1 Introduction

Measurement[1–4] is the key to benchmarking. Comparing the measured performance with world-class performance identifies the gap and so the improvement journey begins[5]. To ensure this measurement is effective, it is important to have a framework on which to base such measurements. It is only by using a common framework that one can reduce the many possible measurements into an effective set[6–7].

2.2 Measurement framework

It is the premise of this book that a world-class process manufacturing plant delivers outstanding customer service[8] from reliable assets exhibiting operational excellence[9–11]. It is operated by highly-motivated people and always maintains its licence to operate by satisfying the high safety and environmental standards of the process industries.

This framework applies to all three aspects of process benchmarking. These are:

- measuring the output of the process which is the performance;
- measuring the input practices that are applied to deliver these outputs;
- measuring the culture and softer issues that ensure the practices are applied and the performance is delivered.

These three categories are discussed in full in this chapter.

2.3 Measuring the performance of a process plant

The framework described above will be used to analyse the individual groups of measures.

2.3.1 Customer service

Process plants produce products that are sold to customers. These may be through pipes direct to the customer's plant, in containers such as road tankers or rail transport, through to individual cans of paint, boxes of pharmaceuticals, strips of steel or packages of food. In every case, there is a customer who pays good money to buy these products from a particular company and plant. Hence, it is the customer who ultimately determines what constitutes outstanding performance. In most cases, the customer is able to buy from any supplier therefore it is a visible measure of customer service on which they judge the supplier. It is also the customer who sets the benchmarking standards.

The following sections show some typical measures that can be used for customer performance.

2.3.1.1 On time in full delivery (OTIF)

This is defined as a percentage of all orders that are delivered on time in full to the customer's premises with no defects in the product or supporting information.

OTIF: The delivery of the product on time in full with no defects in the product or supporting information

'On time' is defined as the time which the customer requested the product to be delivered, not the time the supplier agreed after negotiation. In the most demanding industries this question of 'on time' is increasingly being reduced down to a 15 minute window. For example, the product should be delivered at 11.30 pm (plus or minus 15 minutes) to the customer's premises wherever that may be. If this is not achieved, then customers are beginning to return the product to the supplier and asking them to pay compensation.

'In full' refers not only to the contents being of the correct volume with the correct packaging and meeting specification, but that all the paper work is in order. This is a very demanding measurement, but one which is gaining considerable credibility across the process industries. It is applied not only by the supplier to the final customer, but by the process plant to measure its own suppliers, whether internal or external. Supporting such a measure needs a rigorous recording system either by the plant itself, or by the distribution company if this aspect is out-sourced.

2.3.1.2 Customer complaints

This is defined by the number of complaints received as a percentage of the total orders satisfied:

$$\text{Customer complaints} = \frac{\text{number of complaints} \times 100}{\text{total number of despatches}}$$

It should include all complaints, whether suitably found to be valid or not, because in the customer's mind they are all valid. Again, it requires a positive recording system to measure all the customer's complaints regardless of where they come from. While in the process industries, figures of 1% to 2% are often considered acceptable, the most demanding customers measure complaints in parts per 1,000,000. Hence, 100 complaints per 1,000,000 parts/orders supplied would be approaching world class. For a typical process plant, this may mean that there is less than one complaint per month of any form. Increasingly, this type of measure is leading to customer complaints being forwarded directly to the manufacturing manager.

2.3.1.3 Due date reliability

This measures the percentage of time that the products are actually delivered on the date promised.

$$\text{Due date reliability} = \frac{\text{actual delivery achieved} \times 100}{\text{quoted delivery}}$$

It is a variation of OTIF, but is an alternative that could be used. The due date is based on the quoted lead time by the supplier. Again, in the consumer markets, the measures in this area are extremely demanding. Companies are expected to achieve standards in excess of 99% to be approaching world class.

2.3.1.4 Adherence to production plan

In a world-class company, a sales and operational planning process (S&OP) will exist. A monthly S&OP meeting takes place chaired by the business manager. It has rolling plans from 18 months, three months, one month and will include the possibility of weekly or even daily updates. In the leading applications, the plan is now a real-time plan which is continuously updated by each order received. Adherence to production plan is:

$$\left(1 - \frac{\text{actual production - forecast production}}{\text{forecast production}}\right) \times 100$$

The measure here is normally how the actual production matched the planned production. Such a deviation is normally in the form of a plus or minus %. To get a measure, it is normal to add the absolute value of all the percentage deviations and take the average over the period that is being measured. This is again a very demanding measure, and figures in excess of 95% will be indicative of a manufacturing plant that is well in control and reliable.

2.3.1.5 Stock turn

This is defined as the total sales from the plant divided by the sum of all the stocks involved in the manufacturing process. This includes raw materials, work in progress, engineering spares and so on as well as finished goods. All finished goods between the manufacturing plant and the final customer should be calculated on a consistent basis. If a stock is valued at manufacturing cost then sales must also be at manufacturing cost. Similarly, if final sales value is used then stock is calculated at final sales value:

$$\text{Stock turn} = \frac{\text{annual sales}}{\text{total stock value}}$$

The reasons for holding stock are many and varied. It could be strategic stock in anticipation of some process or plant unreliability or a change in the world supply market; financial reasons where any significant price rises are anticipated and the products have been bought ahead of this; stock may be held due to process unreliability; in anticipation of planned shutdowns or in consignment form to ensure that the customer is never let down. Figure 2.1 is one way of representing the information. Hence, stock turn may be used as a measure of customer service or operational excellence.

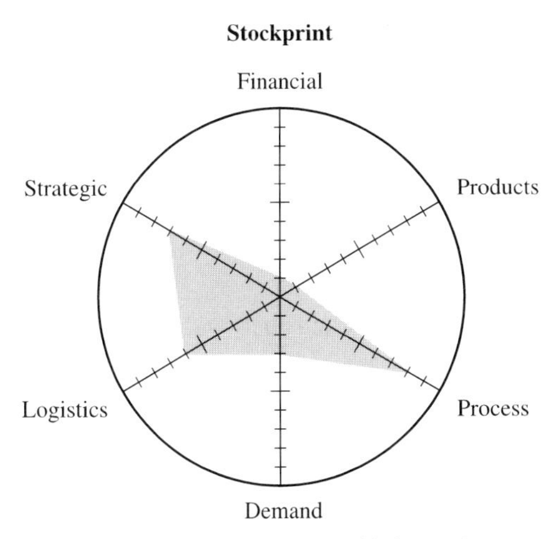

Figure 2.1 The representation of information on stock turn

What is clear, however, is that world-class companies continuously work to reduce stock. Every effort is made to reduce stock holding and to focus on making to order and shipping on the due time rather than from stock.

It is notable in the process industries that very high stock turns tend to occur where the chemicals are hazardous for some reason — for example, ethylene oxide, phosgene and chlorine. In these circumstances, for safety and environmental reasons, it is highly undesirable to maintain stock but it normally proves possible to achieve very high delivery reliabilities without maintaining the stock. This establishes the principle that there is no reason why high stock turns in excess of 25 should not be achieved in the chemical industry. Stock turns tend to be slightly lower for multiproduct plants due to the larger product wheel.

The appropriate combination of measures is determined by the nature of the actual industry. For example, a bulk chemical industry may focus more on the adherence to sales and operational planning (S&OP) whereas a consumer product industry is encouraged by its final customers to pay particular attention to issues such as OTIF and customer complaints.

2.3.2 Reliable assets

Just how reliable are the assets? Do they deliver the outstanding performance expected from a world-class operation?

It is increasingly recognized that to measure this, the process industries are adopting a term from the piece-part industry, namely the overall equipment effectiveness (OEE). This works on the principle that the maximum any process plant can make is if it constantly runs flat out, always produces a perfect product and never breaks down. The closeness to this performance is the calculated product of three individual terms. This is also described as uptime, occupacity and utility though these have slightly different definitions:

$$\text{OEE} = \text{product rate} \times \text{quality rate} \times \%\ \text{right first time} \times \text{availability}$$

2.3.2.1 Product rate

This is the average output the plant delivers when it is operating divided by the maximum it could achieve over the same period:

$$\frac{\text{Average production rate}}{\text{Highest achieved production over a seven day period}}$$

$$= \frac{\text{Good production} + \text{potential production in periods of no demand} + \text{failed QC}}{(8760 \times \text{MPR}) - \text{production lost due to shutdown}}$$

The maximum is calculated as the number of days it operates multiplied by the maximum proven rate (MPR). For any plant of any nature, the MPR is the maximum output that has been achieved for any continuous seven-day period. The basis is that if it can be delivered for one seven-day period, it should be deliverable for every seven-day period. For batch processes, the MPR is determined by evaluating best achieved batch times over the past year for each product. This is not a function of design, nor of the theoretical design limit, but is a direct figure based on the recorded maximum output. Note that it is not affected by periods of no demand.

Many process plants invest heavily in both capital and technology to improve the process efficiency and output of the given plant. The consequence is that the MPR should be continually improving with time. If this is the case, then the product rate efficiency probably needs to be measured on a monthly basis and averaged over the year to produce the correct figure. One result is that the reported OEE may decrease with time. Debates may occur where a plant argues that operating slightly below the MPR improves the reliability. This could be a valid argument provided this reliability is apparent elsewhere in the calculation.

2.3.2.2 Quality rate

This is the percentage of all product that is produced first quality without recycling, reprocessing or reworking:

$$\text{Quality rate} = \frac{\text{good production}}{\text{good production} + \text{failed QC}}$$

For a continuous process, the quality rate is the percentage of all feedstock that is produced on specification first time through the plant in the way the plant was designed. For a continuous plant this may be difficult, since it is possible to increase the reboil or reflux rate of separation and so on and thus influence this factor. This, however, will appear as higher energy costs giving a higher variable cost per tonne.

For a batch plant, it is the percentage of batches that are right first time produced without any adjustments. It may be argued that some plants such as paints or food plants are designed to incorporate adjustments at the end of the batch, but the principle here is that it should be right first time without adjustments.

2.3.2.3 Availability

This is the percentage of the year the plant operated (% of 8760 hours which is the total number of hours per year):

$$\text{Availability} = \frac{8760 - (\text{number of hours of total shutdown})}{8760}$$

If the plant shuts down for holidays or any other reason, this detracts from the overall availability. Unavailability is made up of two components: the schedule shutdowns which are for items such as major overhauls, pressure vessel testing or routine inspection; and the unscheduled downtime which arises from breakdowns.

The one exception is if the plant shuts down due to a lack of raw material or final product sales. This is not a function of plant availability, but of the supply chain issues which are measured and recorded elsewhere. Hence, unavailability of these reasons should be removed from the calculation before the availability is calculated. Similarly, in some plants where the asset is relatively low compared with the working capital and customer service requirements, then the plant may only operate for a percentage of the week — for example, a plant which operates five days a week on two shifts. In these cases, the availability is calculated as a percentage of the time the plant actually operates. It is, however, noted and recorded for benchmarking purposes whether there is a considerable hidden potential in this plant to produce a lot more should the market conditions be influenced in that direction.

The OEE is a product of the three terms outlined above and, as such, is a very demanding parameter. For example, to achieve 95% in each of the three categories will still produce an OEE of 87.5%. As a rule of thumb, world-class OEE for a continuous process plant is 95% and for a batch and downstream plant is closer to 85%. It is a function of the nature of the process and of the chemicals being produced — for example, cleaning liquids will anticipate a higher OEE than a plant handling solids or very viscous materials. Experience suggests that this difference is much less than plant operators might imagine. There is a compounding effect, since a reliable plant will tend to achieve a higher first-pass first-quality yield and tends to operate at a higher product rate.

The benchmarks in this area are not as easy to find. They are a function of the nature of the processes and chemicals being made. It is in this area that companies are increasingly finding opportunities to share benchmarks. World class, however, is not a function of the actual process design or of the type of chemical. For example, if somewhere in the world an industry produces a radically new process or route to manufacturing which achieves a much higher

OEE than the conventional route, then this becomes the world-class figure. It is against this target that all plants are judged.

2.3.2.4 Other measures

There are other measurements which can be used in this area — for example, scrap or yield loss:

$$\text{Scrap loss} = \frac{\text{scrapped production} \times 100}{\text{total production}}$$

This is a percentage of all products that are not sold and have to be scrapped in some way. This is a measure of the efficiency of conversion and would encompass the losses incurred at start-up and shutdown. Some plants refer to these as grade two products, but however measured it is extremely useful.

2.3.2.5 Time spent on rework/reprocessing

This is another measure that may be used to quantify the amount of capacity lost by the plant having to rework an off-spec product:

$$\text{Time spent on rework} = \frac{\text{amount of capacity used for rework} \times 100}{\text{total capacity}}$$

Any rework whether blended or directly processed reduces capacity for the plant and appears in the product rate figure outlined in Section 2.4.

2.3.3 Operational excellence

How can the plant be ensured to operate well? There are a number of potential measures of operational excellence such as:

- units of output per day;
- number of defect-free batches;
- defect rate described as 'six sigma';
- conversion efficiency;
- manufacturing velocity;
- control charts on each key variable;
- process capability.

The purpose of a plant is to make a consistent product therefore the way to measure its performance is to measure this consistency. All products have a specification agreed in some form with the customer and this has upper and lower limits. It is possible to routinely analyse the final product quality parameters and to plot the specifications as a distribution. The performance of this is normally measured by statistical process control.

2.3.3.1 Statistical process control

The terms process specification (*Cpk*) and product specification (*Ppk*) are used to measure how close the actual product specification is to the desired specification. A figure greater than one is good whereas a figure less than one is poor — for example, a figure of 0.5 would indicate that approximately 15% of the product was off-specification needing recycling or reprocessing at some time. Hence, a statistical measure of process quality is the key. The problem arises when a product is made to several specifications and it is suggested here that the average *Cpk* of each specification is plotted which in itself gives a further distribution from which one can determine an average *Cpk* for the product (see Figure 2.2).

It is worth noting that world-class companies measure product consistency and, if this is not measured, then it would be normal to give a very low or zero score. Many plants often use laboratory analyses of every batch or product dispatch to measure consistency. The distribution and cyclical properties of this analysis are often measured. This is not, however, world-class performance, since clearly there are resources involved in the laboratory and materials stored in a 'waiting testing' area which would be unnecessary if the product actually achieved product consistency first time.

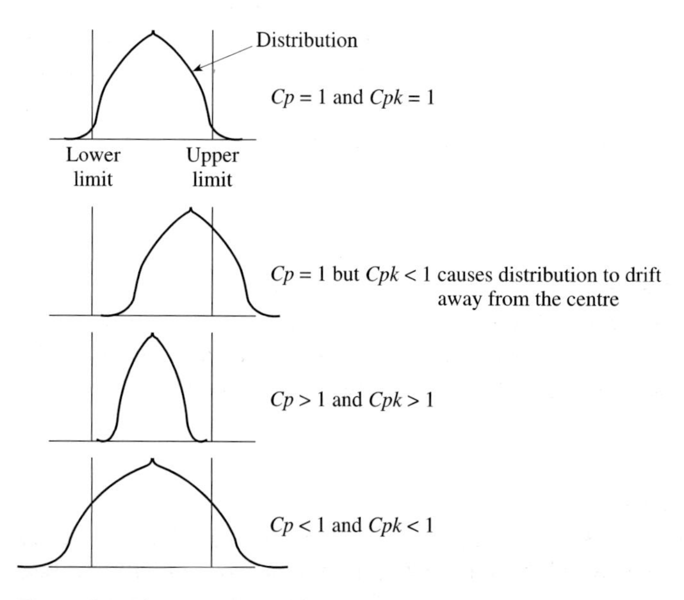

Figure 2.2 The process capability curve

To achieve product consistency, the process is expected to operate at a consistent set of conditions which can be defined precisely and where a statistical band can be put around them. In fact, control charts are often used in world-class companies on the plant to measure the consistency of plant operation and hence deliver the final product consistency.

2.3.3.2 Manufacturing velocity

This measure is widely used but is yet to have a full impact on the process industries. It measures the amount of time that a product has value added to it divided by the total time between raw material arrival and delivery of product to the customer. Manufacturing velocity is:

$$\frac{\text{Amount of time spent on product for added value}}{\text{Total time between raw material arrival and delivery of product to customer}}$$

The value-added time would probably be just the time within the process from inputting raw material to outputting the final product from the plant. For most processes, it is often measured in minutes and even for a large batch process it is a matter of hours. This is compared with a typical process industry's stock turn of seven which is equivalent to 1250 (8760/7) hours. This would suggest that the manufacturing efficiency in the process industries is three or four hours divided by up to 1000 hours so a figure as low as 0.3% is often measured. World-class companies, however, are often aiming to achieve between 5% and 20% in this area. It is a very revealing and demanding measure of the manufacturing efficiency. Ideally, to measure the manufacturing cycle time, the time of each stage should be worked out making the cycle time the sum of the contributions.

2.3.4 Motivated people[12]

Winning companies know that people make the difference. It is winning people who create winning companies. How can the motivation and expertise of the people employed be measured? There are three potential measurements which can be used.

2.3.4.1 Absenteeism

This is measured as the percentage of time all employees in the factory are absent to the total available time including short-term and long-term sickness but excluding any annual or statutory holidays. A typical world-class figure would be less than 1%. Experience supports this is a good measure which is not a function of country, nature of industry or type of behaviour. Crude though it is, it can be a very revealing measure of the motivation of employees.

2.3.4.2 Training days

This is really a measure of the effort that is put into individuals to raise their capability and hence their ability to contribute to ensure winning performance. Again, it should be the number of training days per year for all employees, and it is often appropriate to distinguish between 'on the job' and 'away from the job' training. It is sometimes useful also to differentiate between the training given to new employees and existing ones. Again, there is strong evidence that world-class companies are actively spending in excess of 10 days per employee per year in the whole area of training and up-skilling.

2.3.4.3 Staff turnover

This simply measures the number of people leaving a company's employment as a percentage of the total number of employees. This again is a measure of motivation and it tends to be a low percentage where the company is very competitive and values people. It tends to be very high where the opposite is true, irrespective of the competitive labour market adjacent to any particular manufacturer.

2.3.5 Safety, health and environment

There is now very strong evidence that world-class manufacturing performance leads to world-class safety, health and environmental performance. The evidence points, however, to the fact that the opposite is not true. It is possible to deliver a good safety performance without delivering an outstanding manufacturing performance. This is because the effort is focused on the inherent inputs of safety rather than making it a consequence of achieving manufacturing excellence in all aspects. There is a wide range of performance measures that are used to report safety, some of which are statutory requirements such as:

- Safety:

— fatal accident frequency rate
— 1,000,000 hours per reportable incidents;

- Health:

— days sickness
— illness per 1000 employees;

- Environmental:

— consent violations per year
— total number of leaks.

In developing a set of measurements for any given process industry, as a precursor to benchmarking, it is good practice to go for fewer rather than more measurements. It is more focused and makes the process more effective. As a guideline, it is suggested the total number should probably be less than 15. The above sections have listed over 25 potential benchmarking measurements. In developing the appropriate set for each plant/company, it is important to have measurements in each of these areas:

- customer service;
- reliability;
- operational excellence;
- people;
- safety, health and environment.

It is also important to consider the availability of the information and data.

2.4 Scoring the manufacturing performance[13]

To benchmark, it is necessary to compare the actual performance as measured by some of the above methods against a world-class standard which has been developed and defined from the sources indicated above. The difference between the actual measurement and the world-class benchmark is the gap, as indicated in Table 2.1, page 18.

This performance gap can be converted into a financial gap which is illustrated later in the book. It is appropriate to develop a non-linear scale for each of the measurements. One extreme of the scale could be a very poor performance while the other extreme could be world-class performance. By calibrating this scale from one to 10, it is then possible to put a score against each of the manufacturing performance measures used. Details of this calibration are normally confidential to each user organization and each will attract a score between one and 10. By adding all the scores together it is possible to get an overall score and measure this as a percentage. The purpose of the score is not to produce league tables or comparative performance, but to discover the distribution of the manufacturing performance across a range of plants.

As the purpose of benchmarking is to encourage continuous improvement, it is desirable if the distribution of scores on any group of manufacturing assets improved year on year. This is strongly encouraged by peer site managers sharing results and exploring underlying reasons for differences. It needs to be recognized, however, that world class improves year on year so it is theoretically possible for a plant performance to improve but for the actual score to remain unchanged. This also suggests that world-class performance measurements should be reviewed annually, or at least every two years to ensure progress is being made.

Table 2.1 Measures of the process industries' manufacturing performance

	Average of the worst five plants	Average of the other plants	Average of the best five plants	World class
OTIF %	40	90	99.9	> 99.9%
Customer complaints (% of orders)	6	1	0.01	< 0.01%
Adherence to production schedule (%)	40	85	99	> 99%
Production rate	60	80	99	> 99%
Quality rate	45	90	99	> 99%
Availability (%)	70	85	96	> 97%
OEE	20	60	94	> 95%
Process capability (*Cpk*)	0.6	1	1.5	> 2
Stock turn	4	12	19	> 25
Maximum average time to change a grade (minutes)	480	240	5	< 5 minutes
Absenteeism	10	4	0.8	< 1%

2.5 Summary

This chapter demonstrates the practicality of measuring the manufacturing performance of any process plant. The measurement set produced and used will be influenced by the nature of the industry and the market. There is no standard set but there will always be one which is appropriate to a process business.

References in Chapter 2

1. De Toni, A. and Tonchia, S., 1996, Lean organisation, management by process and performance measurement, *Int J of Operations and Production Management*, 16(2): 221–236.
2. Dixon, J.R., Nanni, A.H. and Vollman, T.E., 1990, *The New Performance Challenge — Measuring Operations for World Class Competition* (Honewood, Illinois, USA).
3. Neely, A.D., Mills, J.F., Platts, K.W., Richards, A.H., Gregory , M.J. and Bourne, M.C.S, 1996, Developing and testing a process for performance measurement system design, *3rd Int Conf of the European Operations Management Association, London, 2–4 June.*
4. Prabhu, V.B. and Yarrow, D.S., 1998, A practice-performance study of North-East manufacturing and service sector industry, *Northern Economic Review*, 27, Spring/Summer.
5. Miller, J.A., 1992, Measuring progress through benchmarking, *CMA Magazine*, 66(4): 37.
6. Voss, C.A., Chisa, V. and Coughlan, P., 1994, Developing and testing benchmarking and self-assessment frameworks in manufacturing, *Int J of Operations and Production Management*, 14(3): 83–100.
7. Cranfield School of Management, 1999, UK Best Factory Entry Form, *Management Today*, http://www.bestpractice.haynet.com.
8. Wallace, T.F., 1993, *Customer Driven Strategy: Winning through Operational Excellence* (John Wiley and Sons, UK).
9. Kaplan, R.S. 1990, *Measures of Manufacturing Excellence* (Harvard Business School, Cambridge, MA, USA).
10. Gardiner, K.M., 1998, Manufacturing processes for the 21st century, preface, *Proceedings of the 12th Conference with Industry, Centre for Manufacturing Systems Engineering, Leigh University, USA, May.*
11. IChemE, 1996, *Process Competitiveness in the 21st Century* (IChemE, UK).
12. Jordan, P., 1998, Steps to improvement — developing people, *Manufacturing Engineer*, 77(5).
13. DTI, June 1995, *Manufacturing Winners — Creating a World-Class Manufacturing Base in the UK.*

Sources of benchmarking data

3

3.1 Introduction

The previous chapter provided guidance on a set of measures which could provide the framework for benchmarking. This chapter assumes that the overall framework has now been selected. It addresses questions such as where to find the benchmark data and suggests some typical figures. In addition, it explores the impact of basic process technology[1–6] on these benchmarks and how different industries affect them.

The basic message is that benchmarking targets are readily available[7–12] once the framework has been agreed upon. They exist and are derived from the literature a process engineer uses and reads, the contacts with customers and suppliers and the personal experience from working on and operating process plants.

World class is an absolute concept. It represents what is the best performance or practice anywhere in the world. In terms of output, it is independent of the nature of the industry or the process. For example, the lowest manufacturing cost per tonne in the world is a fact. If it is derived due to some 'unfair' advantage in the raw materials' price, using old assets, or adjacency to the customer, then that is just a fact. While it may be unfair and may be subsidized, this in no way changes the fact that a particular supplier can supply to the market at a lower price than another supplier. Hence, this becomes a world-class target and other competitors have to try and achieve this target to stay competitive. Similarly, the practices are equally absolute. It is well known in the industry that one or two companies achieve outstanding performance in safety. Such companies set the standard which others must try and attain. To do this, they probably need to adopt similar practices to the best in the world. In benchmarking, world class cannot be changed as it is an absolute.

The accuracy required of world-class performance and practices is influenced by the current performance of the assets being benchmarked. For example, if world class is thought to be around 94% of some parameter, and the actual plant performance is at 70%, then the gap will be 24% and it is really not that important whether the world class is 94% or 95%. If, however, the plant performance is at 94% then it becomes very important to be precisely correct that world class is 95%. As a general observation, the closer any process plant is to world class, the more the total plant team knows what world class is and the greater it perceives the opportunity to improve. The same applies to practices. If a plant has a very poor safety record then it is not necessary to know exactly how world-class performance is achieved in order to improve it. On the other hand, if the plant is very close to world-class safety performance, then the plant team probably already knows what it needs to do to further improve it because it will be paying significant attention to the issue.

Given these basic guidelines, this chapter now explores the source of the benchmarks and the impact of process technology and other industries. Finally, it discusses the appropriateness of having a scoring system.

3.2 Sources of benchmarks

3.2.1 Performance outputs

As has already been explained, performance benchmarks measure the output of business processes. They generally do not take into account differences in the nature of the basic processes. The most frequently-asked questions seem to be what are the sources of benchmarks and where are they available. The general answer is that benchmarks are all around us. By being clear what benchmarks are required, it is possible by surveying published literature[13–15] and openly discussing with customers and suppliers, to determine any benchmark for any parameter quite legally and openly. The following sections give examples.

3.2.2 Customer performance

The most demanding customers determine world class in customer performance. These are the consumers in the most sophisticated markets so it would be natural to look towards the USA, Europe and Japan for those measures. It is in such countries or markets where there is excess capacity and customer choice, that the most outstanding performance can be found. As a whole, customers are unreasonable people, expecting the supplier to deliver a perfect product at a time and place where they want it with absolutely zero defects. Therefore, to determine world-class performance in customer service, one would tend to look towards the retail and automotive markets. It is a feature of these markets that there is extensive benchmarking data published and reported. Perhaps the best example of this is in Womack and Jones[16,17] which benchmarks the whole automotive industry worldwide. Similarly, numerous Governments provide information on the subject such as the UK Best Factory Awards. Case studies from the Best Factory Awards are published in *Management Today* and other magazines and many benchmarking services are available[18–26].

3.2.2.1 On time in full (OTIF)

Typically, world-class OTIF is in excess of 99%. This basically means that an excess of 99% of all orders are delivered on time in full when the customer requires them with no defect in the products or paperwork or any other aspects. The trend is always to improve this and world class is currently approaching 99.7% and one can even envisage it approaching 99.99% in the not too distant future.

The process industries often say that they are not consumer industries therefore their customers do not expect this level of performance. It is, however, world-class performance and some process companies are achieving it particularly in the food area, so what becomes the norm at one end of the supply chain will eventually become the norm throughout the whole supply chain.

3.2.3 Customer complaints

Within any process manufacturing plant, there will be some form of sales or supply organization. They will receive customer complaints whether encouraged or not. World-class companies would actually encourage customer complaints as a source of continuous learning. Hence, the first source of benchmarks is to ask the supply organization what is currently achieved across any particular company for all of its products. The next source is to ask the customers what is the best customer service they receive from suppliers. These will give indicative figures. World class is probably determined by the automotive industry. The first thing to recognize is that they measure customer complaints in parts per 1,000,000 not in percentages. Typically, they are aiming for defects of less than 100 parts per 1,000,000. This would be described as customer complaints of 0.001%. In addition, world-class industries would be encouraging these complaints so these are even more demanding targets.

3.2.4 Delivery due date

The source of benchmarks in this area is via the logistics companies. They will no doubt say that their most demanding customers ask for delivery of their materials within a certain window, probably plus or minus 15 minutes. It is the supplier's responsibility to take into account the conditions of the roads, the weather and so on. Very demanding customers would say that if the material is not delivered within that time slot then it will be returned. Hence, they would anticipate a due date delivery in excess of 99% of the specified time. Again, the source of these measures is to ask the logistics companies and the customers.

3.2.5 Adherence to production plan

This is a slightly more difficult area as nominally the customers do not directly see this — they simply see the results. However, world-class data are readily published by the Oliver White Organization as part of the implementation of MRP2[27]. They state that world-class performance would be in excess of 98% of the planned production and that to achieve MRP2 Class A, a company must be better than ± 2 % of the monthly plan. These set the targets which can be achieved and they are continually updated each time that the MRP2 Class A is updated.

3.2.6 Stock turn

For all companies the stock turn figures are available from the annual reports. For many companies in the process industries, however, this is an aggregate of many plants operating in many different regions. It is therefore not possible to break it down to the individual plant level. Even at this level, however, it is indicative of different performances. It highlights the wide difference between the best and others.

The UK Best Factory Award[18,19,25,26,28,29abc] provides data for the process industries and this would indicate that world class is in the order of 20 for a non-continuous or batch process (downstream) and 30 for a continuous (upstream) process. These are very demanding figures given that a more typical figure for the process industries is around eight. It is worth recognizing that in process industries such as food, where for reasons of hygiene the ability to store stock is very limited, then stock turns in excess of 50 are not uncommon.

3.3 Reliable assets

At first sight this may appear to be a very difficult parameter to obtain benchmarks. It is the one area where the process industries normally argue that it is 'different' to other manufacturing industries. Typical arguments include the plant is old; it handles solids; it is a batch asset or it makes continuous film.

The basic observation is that world class is independent of the nature of the process. As a rule of thumb, world class for a continuous process plant has an overall equipment effectiveness (OEE) in excess of 95% and, for a batch downstream type of plant, the OEE is in excess of 85%.

The sources of this data come from external consultants who have measured this ranging from institutions such as JIPM in Japan to our own experience within the process industries.

As OEE is a product of three factors, how these high figures are achieved does vary. For example, a large continuous plant which does not shut down for three years would have high availability and high product rate but may be let down by a quality rate below world class. In comparison, a batch asset may have a relatively low availability of 90% but always produces at the same product rate and its quality is outstanding. Typically, in the food or pharmaceutical industries, the quality rate, product rate and availability is high. Hence, the sources and world class of OEE are well defined. What is much more difficult to obtain is the individual components of OEE for a particular plant or business. Only with intimate knowledge of a particular industry is it possible to derive the individual components. However, as 95% OEE is a product of availability, product and quality rates, it soon becomes apparent that values in excess of 95% for all components are necessary to achieve a 95% OEE. A small error in world class for availability between 97% and 98% is not all that significant if the present plant's availability performance is 82%. Hence, the lack of precision in the components is not critical in the drive to world-class performance.

3.3.1 Operational excellence

A prime measure of operational excellence is the product quality based on first pass, first quality. Product quality is normally measured in statistical terms as in Chapter 2, Section 2.3.3, page 13. World-class product quality is determined by the customer therefore the first source of the benchmarks is to ask the customer. This is normally an area where they are very forthcoming and willing to co-operate. In addition, it is probable that in the annual negotiation on price, service, contract, partnership, or whatever, they will become ever more demanding in the requirements of product quality as they try to bring all their suppliers up to a common standard. The second source of benchmarks in this area is to actually purchase and dissect the competitor's products. This is quite common in some industries and is used widely as a means of determining the standards. Therefore, world class in product quality is readily available and easily obtained.

3.3.2 Manufacturing cycle time/velocity

This is also a difficult measure to establish benchmarks. It is a measurement of the internal efficiency of a particular process which demands a detailed knowledge of the plant operation — for example, the product wheel. However, for a batch process, it is normally possible to estimate from the fundamental chemistry the manufacturing time of a particular batch. It is possible to estimate the total time between raw material delivery and finished product delivery from the average days of stock. Using this information, the manufacturing cycle time and manufacturing velocity may be estimated. This gives an indication of world class. It does not, however, recognize that a competitor may have come up with a totally different route to manufacture with a radically different cycle time. To monitor this involves checking the patents and the published literature with a mind set on manufacturing cycle times. Some indications can be obtained by asking equipment suppliers. They are normally involved in supplying equipment such as novel washout equipment, changeover automations and so on and, by open but careful questioning, it is possible to determine or estimate the world-class cycle times.

3.3.3 People

World-class measures for the employment and empowerment of people are normally published by Government organizations[30–32] which is public information — for example, groups such as the UK Training Councils publish data on training days per employee and the Department of Employment publishes data on absenteeism. The key is to look at this data and identify what is the best performance and wherever possible to check this with data from other countries. UK Best Factory again publishes data in this area which is readily available[18,19,25,26].

It is our experience that world-class performance tends to be below 1% absenteeism with training days in excess of 10 days per employee per year, irrespective of country or nature of the industry.

3.3.4 Finance

The financial measures of performance are normally easier to obtain. All companies produce annual reports, and company-wide figures are readily available and often provided by the analysts. It is not difficult to obtain comparative data — for example, Table 3.1 (see overleaf) gives the added value per employee for a range of major companies. This is derived from annual reports[33,34].

The difficulty lies in that these measures are usually on a company basis instead of the individual plants. Manufacturing plants add the value while the overheads take it away. Somehow one needs to get to the plant level. This data

Table 3.1 Representative added value per employee for leading companies

Company	£k added value/employee
Intel	244
HP	220
Ford	264
Deere	233
Texaco	964
Exxon	954
Dow	292
DuPont	263
Coca Cola	400
Mobil	880

is more difficult to obtain, but by careful examination of new assets being announced, it is normally possible to estimate the added value for these new assets. Simply asking around the industry at conferences will provide an indicative figure of the employees making it possible to begin to estimate the manufacturing added value per employee. If a particular plant has an added value of £50,000 per employee and the indicative data shows that world class is £200,000 per employee, then the fact a gap of 50% exists in the number of employees is really not that significant in recognizing that the plant is quite a long way from world class. The UK Best Factory data again provides information in this area and for any company with more than one plant provides useful comparative benchmarks.

3.3.5 Safety, health and environment

This is a totally different area due to the pressures of licence to operate, and the trend has increasingly been towards trade associations and countries providing a standard set of metrics and comparative data for companies in the industry — for example, the Chemical Industries Association[35] in the UK provides figures while company annual reports provide performance measures for these areas. It is simply a case of examining the publicly produced information and looking at the world-class targets they have set themselves to derive these figures. As a rule of thumb, with figures such as 1,000,000 working hours per lost time accident, zero environmental issues are becoming the targets.

3.4 Impact of process technology

For many years the process industries have argued they are different, which is one of the reasons why using benchmarks from other industries do not apply. While this is true with regard to the basic chemical processes, it is less true than has been argued. It is a fact that the process industry plants tend to be expensive, difficult and reluctant to start up and shut down due to safety reasons. Unlike other industries such as electronics, it is not possible to restructure the whole manufacturing process over the weekend. Often these assets were designed and built to operate for over 20 years and in that sense they are relatively inflexible. In addition, they historically tend to have built into them considerable amounts of capacity which makes them less dynamic in the process sense. Having said all that, these are still factors which may influence some of the world-class metrics, albeit only minor.

As will be apparent from the previous section, the world-class performance in customer service and people is totally independent of the nature of the process plant. It is set entirely by the customers. Interestingly, the SHE standards in the process industries are invariably higher than that in other industries. This is an example where the nature of the industry, and often the potential impact of any incident, has led to the industry being its own most demanding customer and thus setting the world-class standards.

Having said this, there are two distinctive characteristics that have an influence on what is world-class and benchmark performance. Figure 3.1 illustrates these three processes.

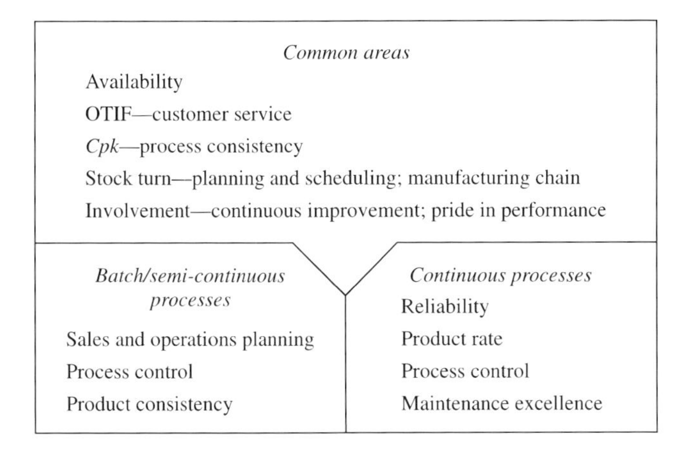

Figure 3.1 The impact of process type on plant priorities

An upstream process makes a very limited number of products which are supplied to very few customers. Typical examples would be olefins, methanol, butane, petrol and so on. They are characterized by being very expensive in fixed capital costs. The issues and measures that drive them are OEE, stock turn, and OTIF.

The downstream plants are the opposite. They are characterized by making many final products and exhibiting a great deal of flexibility. They tend to have a batch manufacturing nature, and the cost of working capital and fixed costs is far more significant than the capital cost of the plants. They also tend to have a more flexible construction. The issues that drive these focus more on customer service measures, lead time, stock turn and possibly the fixed and variable costs.

Finally, there is an intermediate group which is basically in between. It may make in the order of 20–30 products which it supplies in smaller batches of, for example, road tanker size to ever more demanding customers. In this area, a balance is normally struck between fixed and working capital priorities and the set of measures adjusted accordingly.

The point made here is that the important performances and practices to be benchmarked do differ with the nature of the process and industry. What does not differ, is the world-class performance for those industries. This is the major impact for the nature of the process.

It should be noted that basic chemical engineering has not been mentioned. The reason being that this does not make a major difference to world class which always causes a great deal of surprise. For example, if a company is making paint in a large batch reactor which determines the minimum batch size, it will incur some high stock. If, however, world class is determined by a pipeless plant which has a very small batch size and can make to order then this determines world class. The fact the company with the large batch plant would never achieve world class unless it re-invests, is one of the hard realities of world-class manufacturing. Whether they re-invest or do not survive, it does not change world class.

Similarly, if a company has a difficult process involving solids and a competitor comes up with a manufacturing alternative through liquids which is much more reliable, then that sets world class and the company with the solids process is immediately under pressure. A more difficult issue is to obtain a clear understanding of availability benchmarks for different industries. In a large continuous plant, the availability is a percentage of running 24 hours a day every day of the year. But in the food manufacturing industry, the manufacturing cycle may have eight hours of manufacture, three hours of maintenance, and 13 hours of washing and cleaning per day. In these circumstances, what is the availability determined by?

The answer lies in determining what is the most output the plant has ever achieved. For example, if at Christmas a plant manufactures for 16 hours a day, does maintenance for one hour and cleans for seven hours and still meets requirements then the availability becomes the percentage of the 16 hours. The point being made here is that one has to look individually at the manufacturing process to ensure that the targets achieved and being measured are practical.

Another area influenced by the nature of the process is product consistency. If a chemical or food process has not been designed to achieve manufacturing excellence, then it may be difficult to maintain proving impossible to achieve very high product qualities. An example would be a process that depends on the impractical precise mixing of components. In these circumstances, there may be an ultimate value of the product consistency that any process can achieve. That does not, however, change world class. If a competitor with a different process is able to achieve a much higher product quality more economically — that is, by not having to use extensive blending — then that becomes world class and the company with the poor manufacturing process has got a problem. This is the whole benefit of benchmarking.

3.5 Impact of different industries

As mentioned in the last section, the process industries often argue that they are different. The published facts and our experience suggest this is not the case. Tables 3.2–3.5, pages 30–31, are produced from a UK report called *Manufacturing Winners*[18] based on over 1000 UK plants which have entered Britain's Best Factory Awards. These tables plot the average top quartile top 10% performance against a number of measures for four different industries.

The basic observation from this data is that the difference between the average data and the best data in any industry is greater than the differences in performance between industries. As a general rule, world-class performance is evolving to a common set irrespective of the industry. This is particularly true in customer service, people areas[36] and operational excellence.

Two areas differ if, as is often the case in the process industries, the industry has chosen to invest in capital then one would anticipate a much higher manufacturing value per employee than is achieved in a more flexible people-orientated consumer industry such as the food industry. The second area where differences may apply is in the OEE. Again, for example, if a food company at first sight only runs eight hours per day, then it would have a much lower OEE than a continuous plant. However, by ensuring that the factors of availability, first pass first quality and product rate reflect the nature of the industry, one moves to a common set of definitions which increasingly set a common set of targets.

Table 3.2 Typical quantitative benchmarks for the process sector[18]

	Average performance	Top 25%	Top 10%	Your performance
Delivery reliability	88%	97.5%	99%	
Ex-stock availability	86%	98.3%	100%	
New product introduction over last five years (new product/ product range)	2%	7%	55%	
Scrap rate	5.8%	1.3%	0.4%	
Manufacturing added value per manufacturing employee (£000)	72	90	140	
Total stock turns	10	11	19	

Table 3.3 Typical quantitative benchmarks for the household products sector[18]

	Average performance	Top 25%	Top 10%	Your performance
Delivery reliability	94%	98.8%	99.5%	
Ex-stock availability	97%	99.5%	100%	
New product introduction over last five years (new product/ product range)	4%	14%	63%	
Scrap rate	3.9%	1%	0.8%	
Manufacturing added value per manufacturing employee (£000)	45	56	85	
Total stock turns	13	18	30	

Table 3.4 Typical quantitative benchmarks for the engineering sector[18]

	Average performance	Top 25%	Top 10%	Your performance
Delivery reliability	87%	98%	100%	
Ex-stock availability	91%	99%	98.5%	
New product introduction over last five years (new product/ product range)	3%	15%	62%	
Scrap rate	2.7%	0.5%	0.3%	
Manufacturing added value per manufacturing employee (£000)	61	65	90	
Total stock turns	9	11	18	

Table 3.5 Typical quantitative benchmarks for the electronics sector[18]

	Average performance	Top 25%	Top 10%	Your performance
Delivery reliability	85%	98%	99.9%	
Ex-stock availability	88%	97%	98%	
New product introduction over last five years (new product/ product range)	7%	26%	82%	
Scrap rate	3.9%	0.4%	0.2%	
Manufacturing added value per manufacturing employee (£000)	53	62	108	
Total stock turns	8	10	19	

Table 3.6 Typical quantitative benchmarks of performance for different industries[18]

	Engineering	Electronic	Process	Household products
Delivery reliability	100%	99.9%	99%	99.6%
Ex-stock availability	100%	100%	100%	100%
Stock turns	18	11	16	28
Set-up time (minutes)	10	3	10	8
Training days per year	10	10	13	15
Absenteeism	1%	1.3%	1.8%	1.7%

The simple observation is that the nature of the industry is less important than the differences between average and world-class performance (see Table 3.6). Similarly, the practices adopted are becoming increasingly common. As benchmarking gathers pace and as industries choose to benchmark from each other and 'pinch with pride' ideas, then the practices being adopted become more common.

3.6 Summary

This chapter has covered the sources of benchmarks for various measures and comparison with world-class standards. World class is considered an absolute concept which represents what is the best performance or practice anywhere in the world.

References in Chapter 3

1. Boucher, T.O., Jafari, M.A. and Elsayed, E.A. 1994, *Proceedings of Rutger's Conference on Computer Integrated Manufacturing in the Process Industries* (Rutger University, USA).
2. IChemE, 1996, *Process Competitiveness in the 21st Century* (IChemE, UK).
3. IChemE, 1998, *Future Life* (IChemE, UK).
4. Ahmad, M.M., 1998, Global manufacturing competitiveness, *Proc of 8th Int Conf on Flexible Automation and Intelligent Manufacturing* (Begell House Inc, New York, USA, ISBN 1 56700 118 1), pp. 1-13.

5. Ahmad, M.M., 1996, Next generation process manufacturing systems, *Proc of 6th Int Conf on Flexible Automation and Intelligent Manufacturing (Begell House Publishers, New York, USA).*
6. Office of Science and Technology, 1998, Processing the future, *Report of the Foresight Process Industry Group* (DTI, UK).
7. Baldridge, M., 1996, *Winners Profile Booklet* (Malcolm Baldridge National Quality Award Office, National Institute of Science and Technology, Gaithersburg, MD, USA).
8. European Communities Commission, 1993, *White Paper: Growth, Competitiveness, Employment — The Challenges and Ways Forward into the 21st Century.*
9. UK Government, 1994, *White Paper on Competitiveness* (HMSO, UK).
10. DTI, 1994, *Competitiveness — Helping Business to Win* (HMSO, UK).
11. DTI, 1995, *Competitiveness — Helping Smaller Firms* (HMSO, UK).
12. Hanson, P., Voss, C., Blackmore, K. and Oak, B., 1994, *Made in Europe: A Four Nations Best Practice Study* (SBM UK Ltd/London Business School, UK).
13. European Federation of Quality Management — EFQM Model.
14. IMD, 1996, *World Competitiveness Yearbook* (Institute of Management Development, Lausanne, Switzerland).
15. IACOCCA Institute, 1991, *21st Century Manufacturing Enterprise Strategy* (Lehigh University, Bethlehem, USA).
16. Womack, J.P. and Jones, D.T., 1996, *Lean Thinking* (Simon and Schuster Inc, USA).
17. Womack, J.P. and Jones, D.T., 1991, *The Machine that Changed the World* (Macmillan Publishing, UK).
18. UK TEC Manufacturing Council, 1995, *Manufacturing Winners* (HMSO, UK).
19. *Management Today's Guide to Britains Best Factories* — Volumes 1, 2, and 3 (Haymarket Business Publications, UK).
20. Delbridge, R. and Olive, N., 1991, Narrowing the gap? Stockturns in the Japanese and Western car industry, *Int Jnl of Production Research*, 29(10): 2083-2095.
21. Gardiner, K.M., 1993, Globally ideal manufacturing systems: characteristics and requirements, *Third Int Conf on Flexible Automation and Intelligent Manufacturing-FAIM* (CRC Press), pp. 67-77.
22. 1997, Fortune 1000 ranked within industries, *Fortune*, April 28, pp. 44-66.
23. 1976, The 500 largest industrial corporations (ranked by sales), *Fortune*, May, pp. 318–336.
24. 1976, The second 500 largest industrial corporations, *Fortune*, June, pp. 212–314.
25. 1994, *Competitiveness — How the Best UK Companies are Winning* (CBI/DTI, UK).
26. 1995, *Manufacturing Winners — Creating a World Class Manufacturing Base in the UK* (DTI, UK).
27. Oliver White, 1993, *The ABCD Checklist for Operational Excellence*, 4th edn, (John Wiley and Sons).
28. GONE, 1998, *Competing for the 21st Century: A Competitive Strategy for the North East* (Government Office North East, Newcastle, UK).
29a. DTI/CBI, 1997, *Fit for the Future: How Competitive is UK Manufacturing?* (London, UK)

29b. DTI/CBI, 1997, *Competitiveness — a Benchmark for Business* (London, UK).
29c. DTI, 1998, *Our Competitive Future — Building on the Knowledge-Driven Economy* (London, UK).
30. Bureau of Labour Statistics, US Government, various volumes available on the Internet through: http://www.census.gov.
31. Commission for the European Communities, 1995, *Competitiveness and Cohesion: Trends in the Regions — 5th periodic report on the social and economic situation and development of the regions of the community*, OOPEC, Luxembourg).
32. DTI, 1995, *Competitiveness — Forging Ahead* (HMSO, UK). Http://www.hmsoinfo.gov.uk/hmso/document/dti-comp/dti-comp.htm.
33. Gardiner, K.M., 1996, An integrated design strategy for future manufacturing systems, *J of Manufacturing Systems*, 15(1): 52–61.
34. Gardiner, K.M., 1998, Globalization, integration, fractal systems and dichotomies, *Proc of 8th Int Conf on Flexible Automation and Intelligent Manufacturing* (Begell House Inc, New York, USA, ISBN 1 56700 118 1), pp. 15–27.
35. Chemical Industries Association, King's Buildings, Smith Square, London, SW1P 3JJ.
36. Bhattacharyya, K., 1998, Education, training and competitiveness, *Manufacturing Engineer*, 77(5).

Calculating the improvement opportunity

4

4.1 Introduction

The fundamental purpose of process benchmarking[1] is to quantify the potential for improvement and to develop plans for continuous improvement[2–4]. The previous chapters have described a process for both collecting the data and identifying the sources of benchmarks. This chapter focuses on how to quantify the gaps that exist.

While the tools and techniques in this chapter may appear to be very simple, a great deal of judgement is required to interpret the difference between the world-class challenge and the practical opportunity on a particular process plant. All the tools focus on quantifying a gap financially, since it is the financial numbers that drive the priorities and help justify the case for continuous improvement.

4.1.1 Added value per employee

In Chapter 3, one of the parameters calculated was the added value per employee. As a reminder, this is calculated by taking the sales value, minus the fixed and variable costs, and dividing by the number of employees. It is mentioned here because it gives a perspective against which all the following financial gaps may be judged. For example, if the added value per employee is very high, such as £400,000 per employee, then it is unlikely that the priority issue is going to be the number of employees. Similarly, if the value is very low, say £30,000 per employee, then this may say something about the nature of the business and the relative importance of fixed and variable costs. It is always important to have this number in mind as the following calculations are derived.

4.2 Hidden plant

The hidden plant is the amount of extra production that could be achieved from a particular process plant if it was operating at world-class performance. The following shows how to calculate the hidden plant:

$$\text{Hidden plant} = \text{output} \times \left(\frac{\text{world class OEE}}{\text{actual OEE}} - 1\right)$$

World-class OEE: batch > 85%, continuous > 95% unless the business is able to justify a different figure. The assumption is that the sales value directly relates to output volume.

The extra output available will, as a first approximation, be produced at variable cost, and by multiplying the hidden plant by the gross margin, it is possible to estimate the extra profit potential. This profit potential is based on the assumption that every tonne of output can be sold. Clearly, this is a demanding assumption. If the conclusion is the output cannot be sold, then the question arises whether plant capacity should be shut down either permanently or for a prolonged period. This also puts pressure on the commercial side of the business to ensure the sale. There is, however, no argument for purchase-for-resale-type operations for plants which have got hidden plants within them.

At first sight this appears to be a very simple calculation, but care must be taken when interpreting. For example, if the plant being benchmarked is a large continuous plant which operates 168 hours per week, every week of the year and is capital intensive, then the calculation above is correct and as simple as it appears. The challenge is to derive the correct value of world-class OEE, but this is well documented as approaching 95% for such a plant.

The difficulty arises as a plant moves more towards a consumer process plant — for example, paint or food. Such a plant is characterized by making many different products and is probably only in production for a matter of eight to 12 hours per day and for possibly only five days per week. Two difficulties arise.

Firstly, what is world-class OEE? For example, consider a food factory that has the following 24-hour production cycle. From 5 am to 7 pm the raw material is received at the factory. From 7 am to 3 pm the food is produced, from 3 pm to 7 pm the factory is maintained, and from 7 pm through to 5 am the factory is subject to a high-pressure washing regime to ensure cleanliness and sterility.

What is the OEE? What could be achieved? The cycle is logical, and it will always be necessary to carry out high-pressure washing. The answer lies in looking at what is the best that has been achieved in such a factory and it is probable that during peak seasons such as Christmas, it may well have managed to run and produce food for 16 hours a day. This determines what is world-class OEE for such a factory.

The second difficulty is that in a factory of this nature it is probable that the capital cost of the process plant is low and the dominant cost of any extra output is a fixed cost incurred in the labour. To increase output and release in plant would probably demand the employment of extra shifts on overtime so the variable production costs may be a long way from the gross margin.

Both these factors need to be taken into account when working out the value of the hidden plant on a downstream consumer-type process. This does not mean there is not a hidden plant, it could in fact be a very large hidden plant. It may, however, be totally uneconomic to release the hidden plant given the present fixed and variable cost economics of the process. This illustrates the point that interpreting the hidden plant, and any other subsequent measures derived, a great deal of experience is required to come up with the correct opportunity.

4.3 Variable cost

There are three variable costs that are indicative of opportunities. These are the variable cost per tonne leaving the factory, the energy cost per tonne making the product and the delivery cost per tonne to the final customer. To determine these factors, it is important that data collection has included the variable cost, energy cost and distribution costs. Again, the calculation to compare with world class is relatively simple:

$$\text{Variable cost gap} = (\text{actual variable cost} - \text{WC variable cost}) \times \text{output}$$

The challenge is to determine what is world class. This is clearly specific to the product being made but not to the process it is made by. For example, the world-class variable cost for methanol will be different to the world-class methanol cost for the pharmaceuticals industry. However, the world-class variable cost for the methanol plant is not a function of the route that it is made. This is one of the major observations of benchmarking processing costs. World-class variable cost is an absolute. For example, an operator of a particular plant may consider the following are unfair disadvantages because:

- assets are old and the process efficiency is low;
- assets are small compared with world-class size;
- competition has subsidized feedstock;
- the plant is remote from the final customer causing large distribution plant costs;
- competition has skilled labour.

Unfortunately, it is an unfair world and by whatever combination of the above or other factors it is achieved, world class is still the target. Whatever process plant is being benchmarked, it has to compete in a world market with world-class performance that may be achieved with plants having many of the above unfair advantages. This, however, sets the challenge which must be met.

The energy cost per tonne opportunity is indicative of the process efficiency and probably the age of the plant. Data for such calculations are often available from published literature, consultant surveys, and contractors who are offering the latest process plants.

The delivered cost per tonne is similarly calculated as follows:

$$\text{Delivered cost gap} = (\text{actual delivered cost} - \text{WC delivered cost})$$

This benchmarks the efficiency and complexity of the delivery process. For example, a single-stream large plant operating a long way from its major customers could anticipate having a high delivered cost per tonne. Similarly, a plant that produces a great deal of grade two product which incurs a great deal of double handling, such as movement from the plant to the store, store to repacking and then distribution to even final customers could again anticipate having a high delivered cost per tonne. As a rule of thumb, delivery costs are less than 10% of the final cost of the product. Hidden within a high delivery cost per tonne are a whole number of factors which need further analysis. High delivery costs indicate an area of potential improvement.

4.4 Fixed cost

Fixed costs are defined as those which exist irrespective of the output of the plant. They are an input to the process and normally include all the labour costs, plus depreciation and any overheads charged by Headquarters. There is always scope to reduce fixed costs, and the impact of the reduction is normally immediate and felt by the bottom line. The problem exists, however, where winning companies know that winning people make the difference. Remove the people, and there is no resource available to make the improvements in the other areas. Hence, this small section comes with the health warning that it is important this is put in perspective for the fixed cost potential. A process plant has to earn the right to have world-class performance and this can occur through a reduction of fixed costs once the performance has been achieved. The calculations of the gap are as follows:

$$\text{Fixed cost gap} = (\text{actual fixed cost} - \text{WC fixed cost})$$

It is the interpretation of fixed costs that needs care. For example, the maintenance cost per tonne may appear at first sight to be high. If, however, the maintenance costs have been calculated as a percentage of the replacement asset value then one gets a second feel as to what is the real magnitude. Figure 4.1 illustrates an anonymous set of data for a large number of plants which plots the OEE against maintenance costs as a percentage of replacement asset value.

It illustrates in the top left-hand corner what would be considered world-class performance — that is, a high OEE coupled with a competitive maintenance cost. The plant has to earn the right to move into that area by improving OEE and, as a consequence, reducing maintenance costs. Hence, the plant with maintenance costs at 6% of replacement asset value and an OEE of 75% does have the potential to save around 3% of its maintenance cost but only when it has moved the OEE up into the 85% plus performance. It is this concept of earning that is critical to all aspects of fixed costs. Similarly, it could be argued that if a plant is incurring fixed costs to raise its safety, health and environmental performance, then it could simply reduce its fixed costs. But again to maintain its licence to operate it has to earn the right to be there.

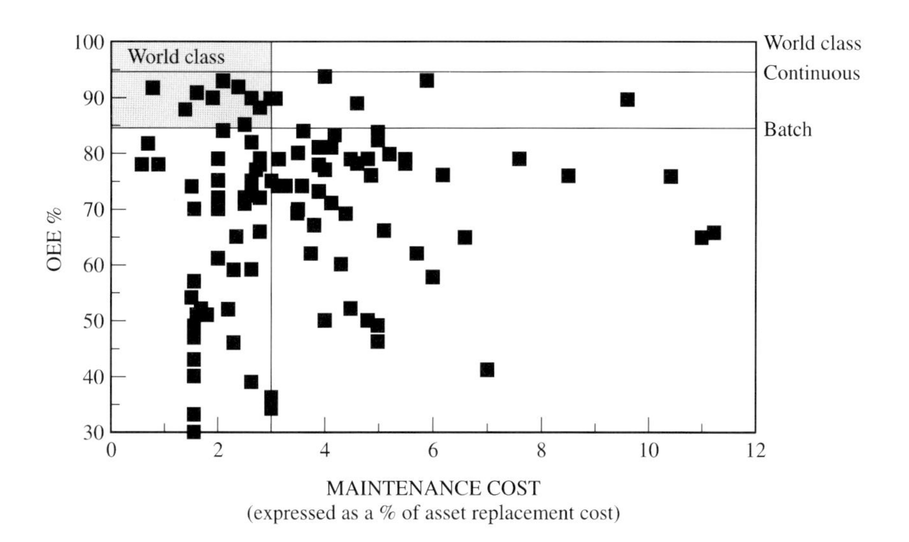

Figure 4.1 A spread of OEE figures plotted against the maintenance spend

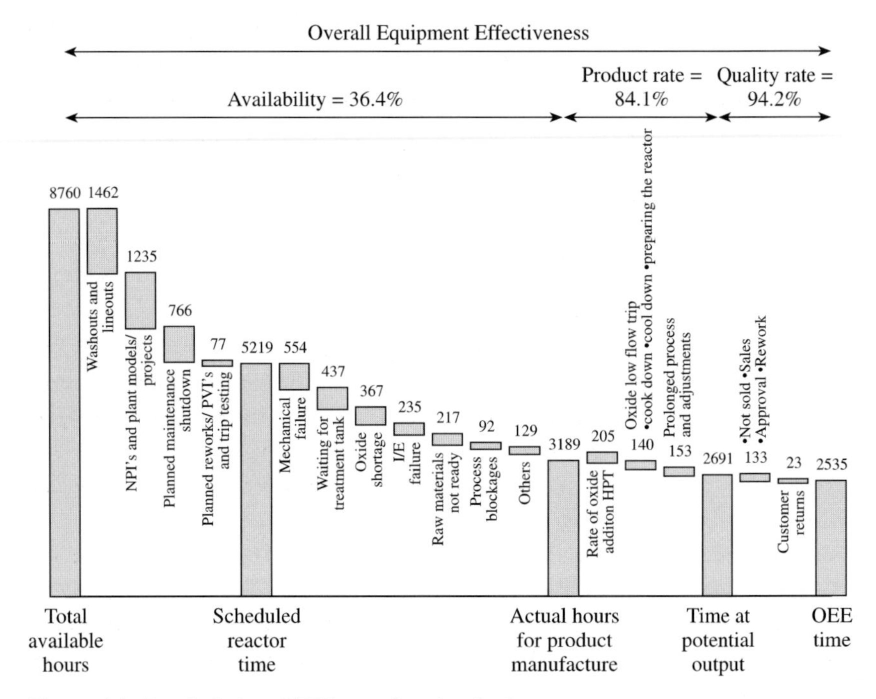

Figure 4.2 Detailed plant OEE losses for a batch plant

If the only data clearly available are maintenance costs, then a rule of thumb is to argue that the fixed cost is twice the maintenance cost and estimate from that. It is a very crude rule of thumb and indicates much further work is required to produce a more detailed analysis. Figure 4.2 illustrates a typical waterfall diagram produced by a more detailed benchmarking process of the fixed costs.

4.5 Stock turn

It has already been said that in our opinion, stock turn is the most useful single measure of manufacturing performance. Achieving a high stock turn demands excellence in all other aspects of process manufacturing.

The opportunity in stock turn is twofold:

- the one-off release of stock[5–8] — this is if the goods can be sold and not replaced in the warehouse and hence not made, then the extra cash released goes straight through to the bottom line;

• there is a potential annual fixed cost saving by not incurring the costs that are involved in storing stock. These would include the cost of warehousing, double handling of materials, handling of returned goods, insurance and funding costs. The figure used by a business varies but, for example, 12% may be a typical figure for the annual cost of stock holding.

The calculation of the potential stock saving is given below:

$$\text{Stock saving gap} = \left(\frac{1}{\text{actual stock turn}} - \frac{1}{\text{WC stock turn}}\right) \times \text{turnover}$$

The challenge in this is again to determine what is world class although this is increasingly well recognized and publicized. In the process industries, plants that produce the most dangerous material such as ethylene oxide and phosgene achieve some of the highest stock turn. These demonstrate that it is possible in the continuous process industries to run very high stock turns and not let the customer down[9].

The stock turn figures are available from annual reports and Figure 4.3 illustrates the stock turn patterns for a number of the major processing industries in the world. These show that while the majority of the companies have maintained a very steady stock turn profile, there are one or two notable exceptions.

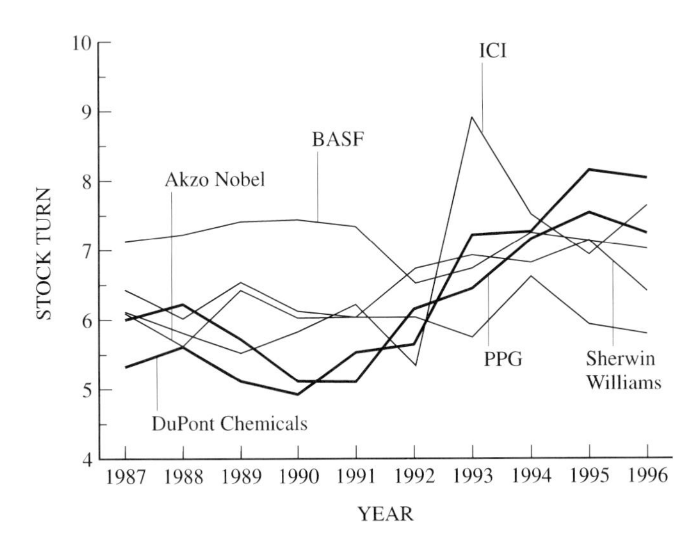

Figure 4.3 Stock turn performance improvement

4.6 Safety, health and environment

This is the one parameter where at first sight it is not possible to obtain a financial opportunity. As stated earlier, world-class standards are well defined hence the gap between the present safety, health or environmental performance and world class is easier to measure and report. The evidence is, however, that there is a strong relationship between world-class SHE and world-class manufacturing performance. In fact, it is world-class manufacturing performance that leads to world-class safety, health and environmental performance[10].

4.7 Summary

In this chapter, the opportunity for hidden plant was identified providing the basis to quantify the potential for the gap that exists. All the tools focus on quantifying a gap as financial numbers set priorities and the justification for continuous improvement.

References in Chapter 4

1. Thor, C.G., 1996, Let's clear up some confusion about benchmarking, *J for Quality and Participation*, 19(4): 26–35.
2. Tenner, A.R. and Detoro, I.J., 1992, *Total Quality Management — Three Steps to Continuous Improvement* (Addison Wesley, UK).
3. Crosby, P.B., 1979, *Quality is Free* (McGraw Hill, USA).
4. Juran, J.M. and Gryna, F.M., 1993, *Quality Planning and Analysis*, 3rd edn (McGraw Hill, USA).
5. Landvater, D.V., 1993, *World Class Production and Inventory Management* (John Wiley and Sons Inc).
6. Mather, H., 1988, *Competitive Manufacturing* (Englewood Cliffe, NJ, Prentice Hall, USA).
7. Wiendahl, H.P., 1994, Management and control of complexity in manufacturing, *Annals of CIRP*, 13(2): 533.
8. Brooks, R.B. and Wilson, L.W., 1993, *Inventory Record: Unleashing the Power of Cycle Counting* (John Wiley and Sons Inc).
9. Wallace, T.F., 1993, *Customer Driven Strategy: Winning through Operational Excellence* (John Wiley and Sons, UK).
10. Benson, R.S, 1997, The link between world class manufacturing and world class safety, health and environment, *Management Today, Manufacturing Excellence*, December.

Signposting the route to process improvement

5

5.1 Introduction

Previous chapters have introduced the concept of benchmarking, suggested possible measures, developed world-class targets and illustrated how to calculate the improvement opportunity gaps from the data. The result of this is a set of quantified opportunities as illustrated in Figure 5.1.

The first task of the person undertaking the benchmarking is to use their experience and mature judgement of improvement processes plus their knowledge of the plant to determine what is practically achievable in the future and what the priorities are for that particular plant. For example, while the hidden plant may represent the largest financial opportunity, the plant operates in a low-cost restricted market so the focus will be to reduce the costs.

Given the selected priorities, the plant is now faced with a plethora of technology and tools that are offered to improve manufacturing performance. The following are just some of the many techniques used: Kanban, Taguchi, Cedac, Kaizan, pinch technology, JIT, SPC, BPR, FMEA, TPM, SMED, poka yoke, agile manufacturing, and flexible automation and intelligent manufacturing[1–7].

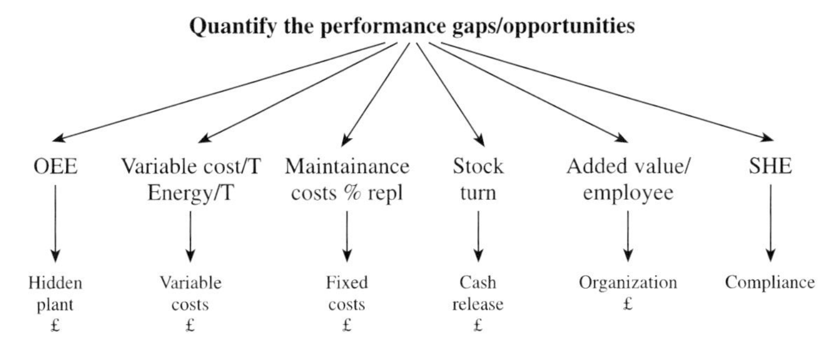

Figure 5.1 Quantify the performance gaps/opportunities

The challenge is to select and use the most appropriate technique. This chapter provides a road map identifying the most appropriate technique for a particular opportunity. It does not go into detail on the techniques since books, courses and papers exist on all of them. In fact, there are probably consultants available who specialize in each one of these techniques. The danger in these circumstances is, however, that it is irrelevant what the problem is as the solution will be the consultant's favourite technique. That is not the approach adopted in this chapter.

The approach is to financially quantify opportunity, quantify the financial benefit of that opportunity and focus on delivering that benefit by 'deep drilling' in the highlighted area. Throughout the chapter, appropriate references are provided to guide the author to more detailed knowledge in a particular technique.

All the techniques in manufacturing focus on removing waste[5,6]. That may not be a familiar concept to all people from the process industries. Waste is anything that does not add value to the manufacturing of the final product. The following lists many of the areas of waste in process manufacturing between the supplier of raw material and delivery to the customer:

- re-processing or blending;
- re-packaging;
- excess reflux rates;
- overtime;
- movement of material between manufacturing stages;
- off-line quality checks;
- storage of inventory;
- unplanned breakdowns;
- relief valve operation.

By reducing or eliminating waste in this context, the product is normally made faster and cheaper. Faster refers to less time between the raw material arriving and the final product being delivered to the customer.

Each of the identified opportunity gaps will be treated separately. Given a particular plant, the reader is recommended to focus on the particular sub-section that is the priority at a particular time.

5.2 The hidden plant (OEE)

All three elements of the OEE calculation, mainly availability, quality rate (right first time) and product rate provide information on which to select the priorities. Figure 5.2 gives a logic for the next level of the road map.

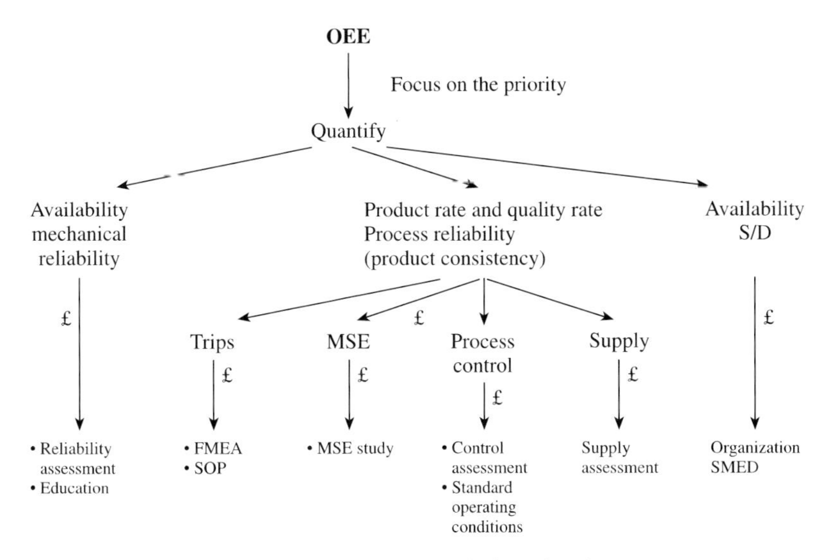

Figure 5.2 Priority areas (reproduced by permission of ICI)

5.2.1 Availability

Availability normally measures the mechanical reliability[8,9] of a particular plant. It is possible to compare the plant's actual availability with either a world-class figure or, if unavailable, the best figure known to the assessor. This identifies the availability gap which again can be converted into financial figures. The calculation is given below:

$$\text{Losses due to availability} = (\text{actual output} + \text{hidden plant})\left(1 - \frac{\text{availability}}{\text{WCA}}\right)$$

where WCA is the world-class availability for plants making the same product.

But this in itself does not tell the whole story. Is the lack of availability due to non-scheduled or scheduled activities? Non-scheduled activities include breakdowns of equipment, trips and so on. Scheduled activities include any major overhauls, statutory inspections, routine testing or routine maintenance of a preventative type. As a rule of thumb, world-class reliability demands that scheduled activities exceed the unscheduled activities and the sum of the two will approach world-class levels. This indicates the mechanical aspects of the plant are in the control, preventative rather than reactive maintenance is practised and the plant is predictable[10].

The tools used include reliability audits, failure mode effect analysis to identify the root causes, predictive maintenance techniques and the replacement of unreliable equipment.

If the scheduled downtime is large, the most normal components are the major overhauls. Particularly on continuous plants, the characteristic is for the plant to have a major overhaul at routine intervals. The theory is that by planning and compressing all the maintenance into a prescribed period, the plant will achieve long uninterrupted runs between these overhauls and the actual shutdown will be more efficient. This may or may not be the case. The whole concept of major shutdowns is contrary to world-class manufacturing principles[11]. A useful indicative tool of the significance of shutdowns is to take the total length of the major overhauls plus the frequency between overhauls and calculate what the consequential loss of that overhaul is in minutes per day. Typically, figures will vary from 20 minutes to over two hours which is quite a significant gap.

The techniques for shortening the length and frequency to the length of the shutdown by increasing the frequency fall in two categories. First and most effective is to remove the need for shutdown at all by challenging the basic premises. These are aspects such as testing relief valves or pressure vessels. With the advent of non-destructive testing, it is now quite possible to double or even triple the time between testing. The second aspect is to reduce the length of the shutdown itself. This is an area where the principles of SMED (Single Minute Exchange of Dies) techniques developed from the automotive industry can be applied. Basically, these techniques identify all the elements of the shutdown then asks what can be taken outside of the critical path as pre-preparation so that the total time of the shutdown is reduced. The experience shows SMED can be very effective in potentially halving the length of a shutdown.

5.2.2 Product rate and quality rate

These two are grouped together since they often are indicative of the same issues. The first step is to compare the product of the two with world-class performance. The comparison lies with the world-class figure which is expected to be in excess of 95%. This gap suggests that this particular plant does not run at the maximum rate and quality, despite the fact that it is not making a product of 100% quality. In other words, the process is unreliable. The calculation is given below:

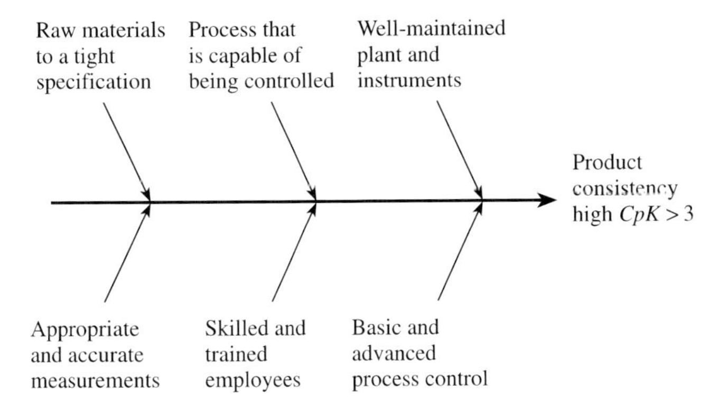

Figure 5.3 Factors which affect overall plant control/consistency

$$\text{Losses due to output restrictions} = (\text{actual output} + \text{hidden plant})\left(1 - \frac{\text{product rate}}{\text{WCP}}\right)$$

where WCP is the world-class product rate for a plant making a similar product.

The symptom for this is a lack of product consistency as illustrated by the poor *Cpk* measures and the need to recycle or reblend products to maintain the final product quality. There are many components which can contribute to poor product consistency. The fishbone diagram in Figure 5.3 illustrates some of the main contributors. These are discussed in detail below.

5.2.3 Supplier quality

A somewhat surprising observation from the process industries is that unlike other industries, they tend to accept much more variable feedstock quality. It is not unusual to visit process plants that have a feed purification stage prior to the actual reaction or conversion operation. This tends not to happen in world-class manufacturers. The onus is on the supplier to deliver the product OTIF to specification and a full customer measurement system is in place to ensure this happens — for example, the measurement of raw material delivery reliability and product *Cpk*. Removing disturbances on the supply side is a key task of the supply chain and the tools and techniques for this are detailed on the Supply Chain Council's website[12]. Removal involves partnership between the supplier and the customer, rather than the confrontational position taken by many buyers.

5.2.4 Process control

This is more recognized for improving product consistency[13,14]. It is, however, necessary to be sure that the process is capable of delivering the product consistency before determining the control improvement required. This is well explained in the book by Shunta[15]. Figure 5.4 illustrates the differences between *Cpk* and *Ppk*.

It illustrates thc differences between *Ppk* and *Cpk*. The *Ppk* measures what the process is capable of achieving while the *Cpk* measures the consistency of the process control. Again, the metrics taken from Shunta makes the point about the differences between *Cpk* and *Ppk* — if the process is capable of achieving the product consistency requirements, then no amount of process control will ever compensate. If the process is capable of achieving the product consistency, then it is appropriate to apply process control.

5.2.5 Trips

A trip is anything that moves the process from its normal operating conditions to a shutdown or within the safe operating envelope. Where a process is under control, trips are extremely infrequent and only arise from an abnormal occurrence such as a total power failure. Unfortunately, this is often not the case on the process plant, and trips always require investigation using techniques such as a failure mode effect analysis that identify the root cause of the trips and removes the problem at source.

This is a very brief introduction to the many aspects of mechanical and process reliability and more detailed accounts are available on this very important topic of what is often the largest opportunity of a particular plant[16–19].

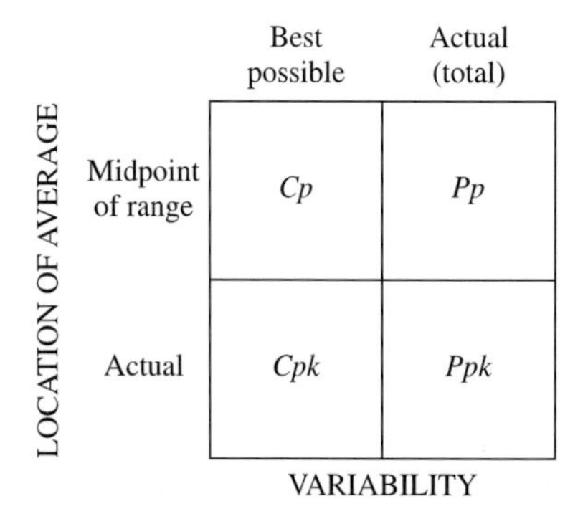

Figure 5.4 Process capability

5.3 Variable costs

The variable cost is a measure of how efficiently the raw materials are converted into finished goods by the process. Figure 5.5 gives a potential root map to analyse the variable cost. This is one of the areas where process engineering has a considerable contribution.

5.3.1 Energy costs

The energy cost per tonne is indicative of scope for improvement. By comparing it with world-class energy costs per tonne, it is possible to calculate the gap and hence the opportunity. This simple calculation is given in the following:

$$\text{Energy gap} = \frac{\text{energy}}{\text{unit of production}} - \frac{\text{WCE}}{\text{unit of production}}$$

where WCE is world-class energy for a plant manufacturing the same product. This gap is typically expressed as GJ/unit of production.

The challenge is to determine what is world-class energy costs. Unfortunately, energy prices are subject to government policy in various countries and it is possible for a plant to have a highly subsidized energy cost. This, however, is reality and a figure must be used to determine world class. If, through a regulated energy price, a competitor is able to produce a chemical at a cheaper price, and sell it on the world market then that determines the competition. A second check is the energy units/tonne of product which will identify if the issue is one of usage or price.

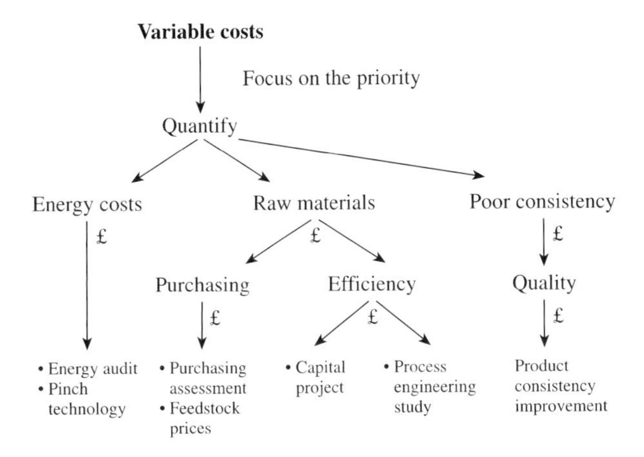

Figure 5.5 Variable costs (reproduced by permission of ICI)

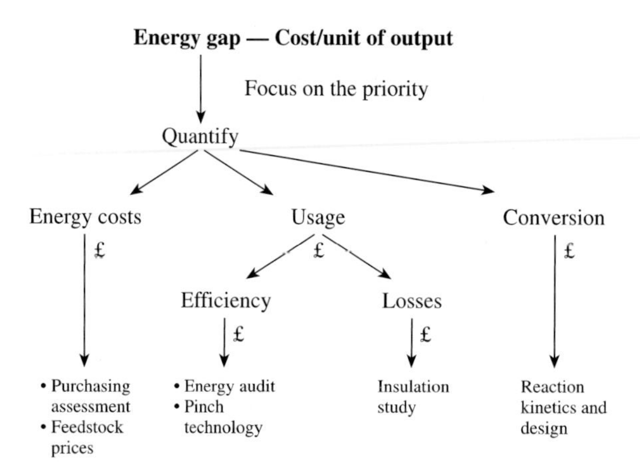

Figure 5.6 Energy gap (reproduced by permission of ICI)

There are many techniques for first analysing and then improving the energy performance. By starting with an energy audit, it is possible to get to many of the root causes. Figure 5.6 indicates some of the paths that might be followed.

The opportunity could range from one of simply improving the lagging, or redesigning the process to maximize the energy interchange using pinch technology. It may, however, require a re-negotiation of the energy prices paid by the plant. This can range from straight negotiation through to the sharing of purchasing among the different number of plants to gain leverage. In the extreme, it could involve appealing to governments to change the tax laws to provide a competitive energy position. The key message is the need to firstly be aware of what is world class so that all of the above techniques are based on factual information.

5.3.2 Raw material purchasing

An extension of the energy cost is the price paid for all raw materials. Again, it is a means of comparing with world-class figures which are often available or may be determined by seeking competitive tender. By applying the rigorous principles of supply chain management, it is possible to determine preferential purchasing arrangements[20]. Many techniques have been written on this subject and Figure 5.7 illustrates some of the options. The raw material purchasing should not sacrifice the incoming quality for marginal cost savings.

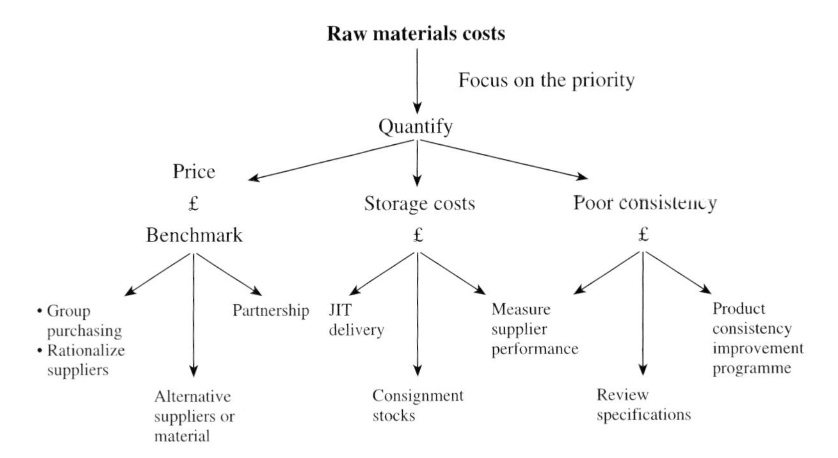

Figure 5.7 Raw material costs (reproduced by permission of ICI)

The trend in world-class manufacturers is to radically rationalize a number of suppliers as a means of reducing the variable cost of raw materials.

5.3.3 Process efficiency

This is heartland process engineering. It is often measured as the conversion efficiency from raw material to final product. Any improvement in the conversion rate of a reaction process not only reduces the amount of waste and environmental impact that is produced directly, but probably reduces the cost of re-processing or separation and hence further reduces the variable cost of production.

5.3.4 Reaction stage

There are many techniques in the full scope of process engineering such as separation, reaction kinetics, process control[21–23] and so on which can be used to improve the variable cost of manufacture[24]. From the manufacturing point of view, they divide into those parts of the process which involve reaction and those which involve physical processing. Figure 5.8 (page 52) gives a road map for the reaction stages.

The key aspects are about catalysis, control and a thorough understanding of the reaction kinetics. Years of experience in industry have shown that the potential in this area is significant but it demands in-depth knowledge.

From the manufacturing point of view, an equally important practice is mixing as it occurs in the manufacture of paint, cosmetics, and other fluids and solids. Good practice is to remove any aspect of re-processing.

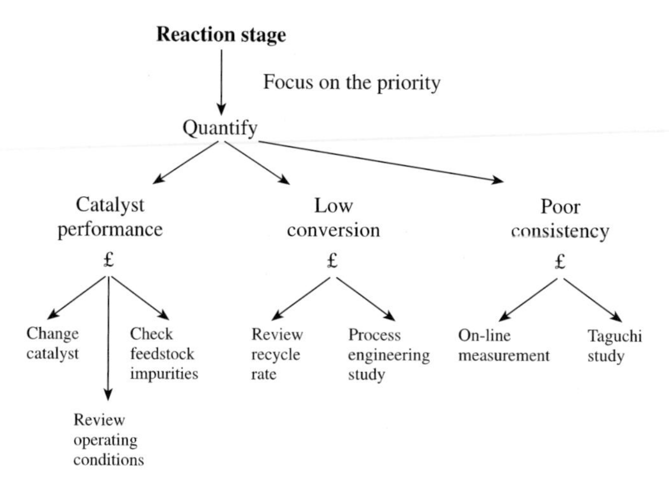

Figure 5.8 Elements of the reactions stage (reproduced by permission of ICI)

Finally, an alternative route to reduce invariable costs is to build a totally new plant which exploits the latest process and manufacturing technology. In doing this, it needs to be recognized that while it may decrease the variable cost, it will directly increase the fixed cost by virtue of funding the capital to expand.

Already, the impact of benchmarking the processes against world-class manufacturer's standards has had an impact on process design. This is described in more detail in Chapter 10.

5.4 Supply chain (stock turn)

The supply chain includes all aspects from raw material to final delivery to the customer. Figure 5.9 is useful to illustrate this aspect. Supply of raw materials and delivery of the product to the customer are elements of fixed costs. It is not unusual in the process industries for the fixed costs associated with supply and delivery to exceed the fixed costs of manufacture. While the next section focuses on the fixed costs of manufacture, this section focuses on the contribution of the supply and distribution cost logistics.

Stock turn is a measure of the output of all aspects of the supply chain. Poor logistics, manufacturing or distribution all manifest themselves in a supply chain that has a low stock turn. Improving supply chain performance by reducing stock not only releases cash and its associated funding, but it results in the manufacturing process moving at a faster speed allowing the manufacturing wheel to turn faster. This removes waste throughout the whole chain.

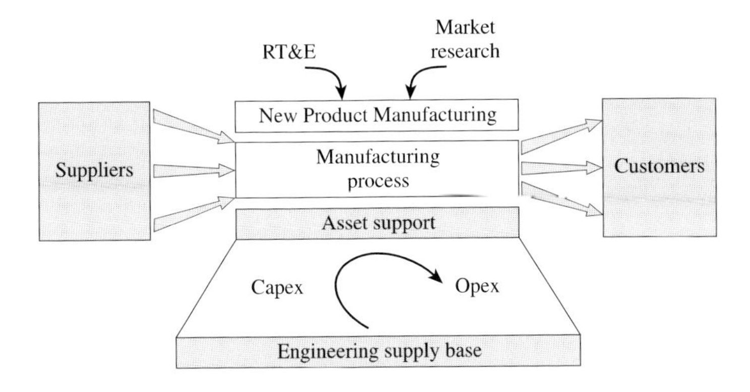

Figure 5.9 Business process model (reproduced by permission of ICI)

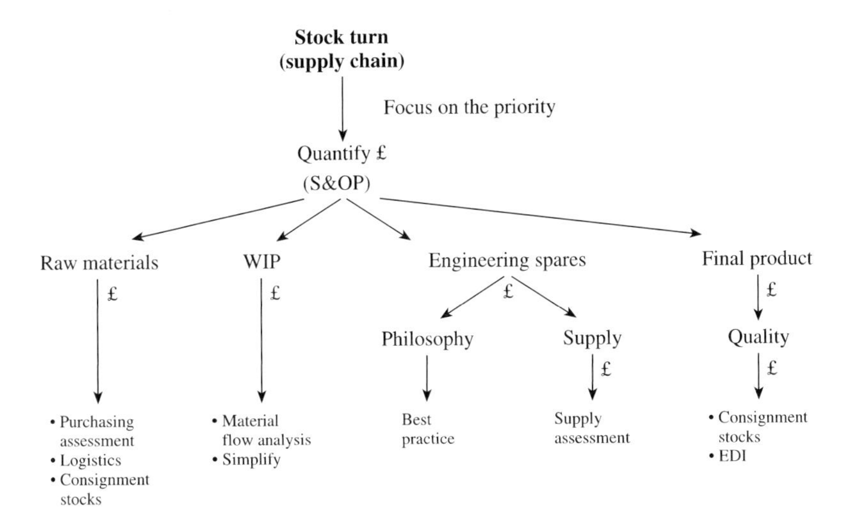

Figure 5.10 Stock turn (reproduced by permission of ICI)

Figure 5.10 illustrates the route map to be explored in improving the supply chain. These are to be compared with world-class performance and are typical of that which might be expected.

To fully understand the supply chain it will probably be necessary to gather additional benchmarking data, particularly in the areas of raw material delivery and finished goods. Tables 5.1, 5.2, 5.3 and 5.4 (page 54) illustrate some of the typical measures to collect in exploring the supply chain options. Definitions are included in Appendix A.1, page 127.

Table 5.1 Benchmarks for customer service

Supply chain	Median	Top 10%
Balance and cover	eight weeks	three weeks
Customer satisfaction	99.0%	99.99%
Customer OTIF	95%	99.9%
Lead time	14 days	one day

Table 5.2 Benchmarks for asset effectiveness: metric values

Supply chain	Median	Top 10%
Process capability (*Cpk*)	1	> 3
OEE		
continuous	80%	> 95%
batch	70%	> 85%
Value added productivity	£100k	> £400k
Grade change times	28 minutes	five minutes

Table 5.3 Benchmarks for supplier performance: metric values

Supply chain	Median	Top 10%
Supplier OTIF	95%	99.9%
Supplier *Cpk*	1	> 3
Raw material stock turn	13	52
Customer leadtime	14 days	one day

Table 5.4 Benchmarks for business processes: metric values

Supply chain	Median	Top 10%
Activity based costing allocation	20%	80%
Supply chain costs	9.9%	6.4%
Cash to cash	77 days	29 days
Forecast accuracy	70%	95%
New product introduction time	24 months	six months

(Tables 5.1 to 5.4 are reproduced by permission of ICI)

These have been derived from the SCOR model[12]. A feature of supply chain is it is not fundamentally different in other industry sectors and therefore benchmark information is readily available from the more widely published industries such as automotive and food.

5.5 Raw materials supply

The tendency in world-class manufacturing companies is to first reduce the number of suppliers, and then develop a partnership with the suppliers who deliver the raw materials just in time (JIT). The implication of this is so both the supplier and the user have a reliable process where it is possible to predict the requirements for raw materials knowing the demand for the product and to share the information openly. Typically, information systems such as MRP2 coupled with the sales and operations plan would be required to make this happen. In addition, it is increasingly common to have some form of direct electronic communication between the supplier and the user to ensure these deliveries.

The benefit from the supply chain is that the user stores minimal stock. The onus is on the supplier to deliver the material when it is required and in the correct quantity, and increasingly if the quality is wrong then the supplier has to take responsibility for the off-spec product. The benefit to suppliers is predictability of demand and reduction of costs and waste. This is a long way from much of the behaviour in the process industries, but illustrates the potential opportunities to be had by removing waste. The user does not store raw materials, and does not need land, building, warehouses, insurance and so on to store it nor all the handling aspects of both receiving the raw materials and feeding them forward to the actual plant. In many parts of the process industries, the raw materials arrive by pipe from elsewhere on the site and in those circumstances there would seem little reason to store any raw materials.

The more normal reason given is because of unreliability of the supplier's plant which is yet again a reason for moving to world-class manufacturing practices throughout the chain. As a point of information, unreliability of raw material supply is not built into the OEE calculation. An extension of this concept of partnership is to move to consignment of stock, where the supplier takes over responsibility for ensuring that the user never runs out of raw material.

5.5.1 Work in progress

This is defined as all the raw material and finished goods that are being used by the actual plant at a particular time. It includes intermediate products in the plant and also includes raw materials and goods being stored on pallets and so on for use in the future. High work in progress is normally indicative of either poor materials flow, excessive process hold-ups or an unreliable process. Unreliability has been covered in Section 5.2.1 and this section focuses on material flow[25,26]. The simple principle is material should flow logically from equipment to equipment with the minimum physical movement between one piece of equipment and the next. Wherever possible, double handling should be avoided and this is referred to as continuous flow. Typically, this is an area where team work and agile manufacture has been developed to minimize the movement of raw materials and the consequential improvement in the speed of manufacture.

5.5.2 Engineering and equipment spares

It is often surprising when completing assessments to discover the extent of the engineering spares carried on a typical process plant. Arguments range from the equipment is always available because the cost of shutdowns is so high, through to the fact there is a special compressor with a drive thus a spare must be carried at all times. Often, questioning will reveal that these spares remain unused for many years and really are there only as insurance. This immediately gives the clue that it may be possible to take out insurance to cover the option and hence not carry the spares in the first place. Basically, the same principles apply here as to raw materials, which is to develop a partnership with the suppliers so that they manage the storage and the delivery of the spares to the plant on time and in full. It is normally possible to negotiate such a partnership, since they deal with many companies all of which have got similar issues enabling the supplier to bring together the economies of scale, systems and logistics to share costs on both sides.

5.5.3 Finished goods storage

A supply chain appraisal will help to understand why the finished goods storage exists. It may be possible to plot it on a diagram as in Figure 5.11.

This identifies what the real reasons are for carrying the finished goods stock. Is it the unreliability of assets hence the lack of confidence, or does the customer insist? Whatever the reason, it is always possible to significantly reduce the amount of stock. Leading competitors turn the finished goods stock into a competitive advantage. This is achieved by considering options such as consignment stock where the supplier takes responsibility for ensuring the company customer never runs out of the raw material, which has many advantages to all parties. It strengthens the partnership with the customer and the degree of dependence. For the supplier, it moves the storage tanks from the supplier's plants to the customer's plants and allows the supplier to use the

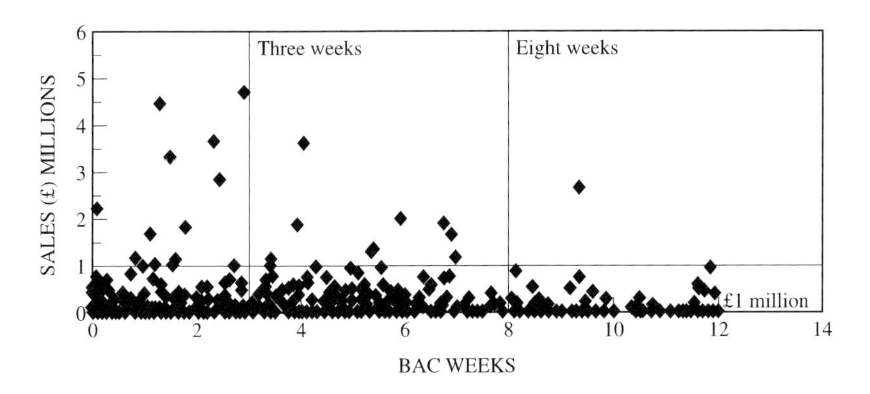

Figure 5.11 Balance and cover for an international lights business

many customers as buffer storage to smooth out any variations in manufacturing performance. It also ensures a much closer understanding of the customer's needs which allows the forecasting of the sales and operations plan accuracy to improve. These are all techniques which help improve manufacturing performance.

5.5.4 Logistics/distribution

Its important to differentiate between the cost of distribution and the management of distribution. The tendency is to out-source distribution to professional logistics firms who use excellent planning techniques to ensure all vehicles are fully loaded to minimize the cost. This is to be encouraged. It does, however, mean that the interface between the supplier and the customer becomes an intermediate third party. Aspects such as customer complaints may move out from the manufacturer to the distributor. This can have serious implications for customer relations. In addition, the lack of direct interface with the customer precludes the options of partnership sourcing and so on. World-class manufacturers increasingly retain control of the management of distribution and the interface with the customers while out-sourcing the logistics to lower the cost.

5.6 Fixed costs

Unfortunately, in many benchmarking studies, it would appear the prime aim would simply mean to reduce the fixed cost. Headings such as lean manufacture have been used to justify this process. The measure used here, however, is the added value per manufacturing employee. This is defined as:

$$\frac{\text{Total sales value} - \text{variable costs} - \text{fixed costs}}{\text{Total manufacturing employees}}$$

Note the use of sales value to the final customer, not the manufacturing price exit the plant. The number of employees is calculated as described in Table A1.1, page 128.

A lower fixed cost per tonne can be achieved by either reducing the fixed costs or by increasing the output. Again, the first step is to benchmark the actual fixed cost against world cost. Where these are not available, they may need to be estimated.

Having identified the areas of fixed costs potential, a route map is required to point to the relevant issues (see Figure 5.12).

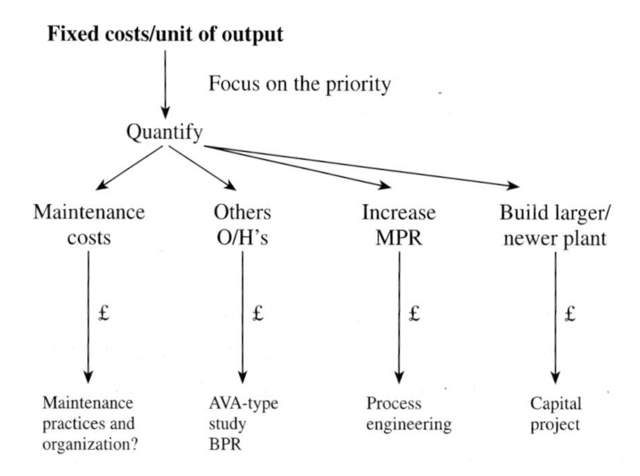

Figure 5.12 Fixed costs per unit of output (reproduced by permission of ICI)

5.6.1 Increase in the maximum proven rate (MPR)

This is normally the most cost-effective route to reducing the fixed cost per tonne. The gearing is such that a 1% or 2% increase in output will have a dramatic effect as the fixed cost remains unchanged. To improve the MPR may require some form of process engineering and possibly some minor expenditure on equipment. In batch processes, this is achieved by reducing batch cycle time. World-class performance in this area is 12% per year compound for increase in MPR which is an extremely demanding figure. It does, however, illustrate the high leverage that is possible by the judicious application of all the technology tools of chemical engineering such as design, kinetics, process modelling and control. It is an area where short-term savings in the number of people employed may reduce the ability to deliver the potential MPR improvements.

5.6.2 Build a larger plant

This again is an approach that has been adopted, particularly in the 1980s, and is based on the belief of economies of scale. This states that as a larger plant requires the same costs to operate as a smaller plant, the fixed cost per tonne is lower the larger the plant is. The limitation is the large capital cost involved plus the risk that the market may change and reduce prior to plant commissioning. While it normally reduces the fixed cost per tonne at the boundary of the plant, it may increase the fixed cost per tonne to the customer when the extra distribution costs are allocated.

5.6.3 Other fixed costs

Before moving into a detailed analysis of fixed costs, it is normally important to identify all the fixed costs associated with the plant, not just the operating costs. These could include the headquarter, technical and research costs and some elements of supply and distribution costs. It is useful to put these into a common framework of units/tonne to give a clear perspective before completing a detailed analysis.

5.6.4 Maintenance cost

These may be a significant proportion of the plant costs. World-class maintenance costs would typically be around 3% of replacement asset value. Notice it is the replacement asset value that is used and achieving this figure has to be earned. It is a challenge to deliver a world-class OEE performance with a world-class maintenance cost. Figure 5.13 (see overleaf) illustrates the distribution of such performance for a large number of plants that have been benchmarked in the past.

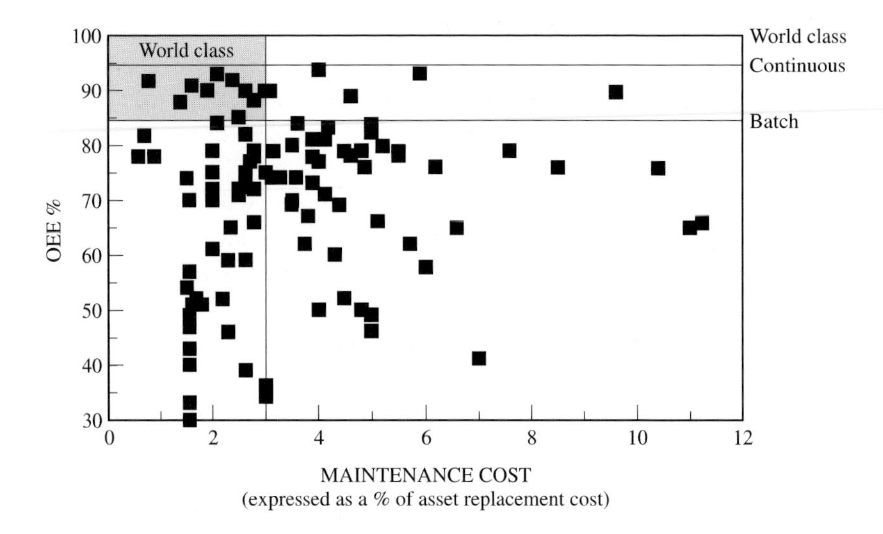

Figure 5.13 OEE versus maintenance costs

This clearly illustrates that the right must be earned to have world-class OEE cover with world-class maintenance expenditure. High expenditure does not necessarily deliver world-class OEE and it is possible to cut the maintenance too far and hence detract from the world-class OEE. Fundamentally, OEE increases normally have higher priority than maintenance cost reduction. It is a fine balance. Earning means that the performance has to be delivered before the costs are cut.

5.6.5 Other overheads

External consultants have a long proven track record of reducing fixed costs in this area using tools such as activity value analysis (AVA)[27] and activity-based costing[28–31]. These work on the principle that one should only pay for the things that are directly associated and bring added value to the plant and products. While this often shows immediate savings, it runs the risk that the longer term aspects of technology or networking tend to be lost and hence the potential to improve the more important MPR may disappear in the process.

In conclusion, it is emphasized that experience shows the best way of reducing the fixed costs is to improve the MPR.

5.7 Organizational effectiveness

Providing a clear figure in this area is a very difficult prospect and all managers have an opinion on the subject. It is suggested that if the manufacturing added value per employee has been benchmarked then it is possible to get a feel for the appropriate number of employees. The logic is as follows.

Firstly, define what the world-class OEE, stock turn, fixed cost and variable cost would be for the plant. From this, determine the resulting added value if at world-class performance and hence calculate the added value per employee when world-class had been achieved. If this is still less than the world-class added value for employee for a plant of that type, then it is probable that the numbers of people on the plant are too large.

This calculation should be done when world class has been determined, not independent of that. It is always possible to cut numbers, but one may well be cutting them below world-class performance with a resultant long-term detrimental damage. The well-known techniques for reducing the organization size are indicated below[32]:

- downsizing;
- rightsizing;
- activity value analysis;
- technical opportunity programme;
- restructuring.

5.7.1 Empowerment/engagement

This is undoubtedly the most effective way. Empowering the employees through knowledge and experience releases their capability to improve their performance dramatically. It may lead to cellular manufacture[33] and other organizational structures but the effect is to release the creative energy of the production and management employees with great benefits. The resulting savings in the fixed costs often occur in the managerial levels which moves from one of being based on information to one that is based on knowledge and experience. This is well described in the characteristics of the winning companies and is referred to as inverting the organizational pyramid.

5.7.2 Other routes

The other routes either occur by increasing the degree of automation, out-sourcing some of the activity or simply downsizing the number of employees. It is critically important for the best performance to ensure that this is done against measurable performance metrics. This will allow one to see if the reduction has been achieved. To achieve it effectively normally involves a great clarity of the organizational development and structure.

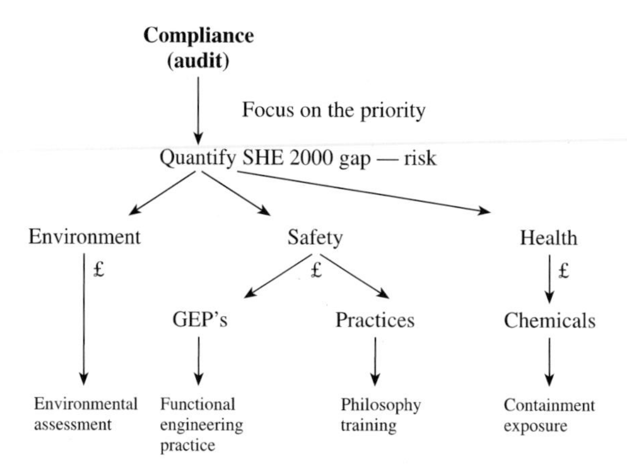

Figure 5.14 Safety compliance (reproduced by permission of ICI)

5.8 Regulatory compliance

This is one area which at first sight would appear not to have a financial benefit. However, a clear relationship exists between world-class manufacturing performance and world-class SHE performance.

This will not surprise anybody in the process industries who recognizes a clean and tidy plant that is organized and well maintained is less likely to have safety, health and environmental issues compared with an untidy plant with tripping hazards, leaks and so on. In fact, it is believed that world-class manufacturing standards are a precursor to world-class SHE standards and to attempt to achieve one without the other would not be sustainable. Figure 5.14 illustrates the route map for improvements in this area.

Principles such as good manufacturing practice[34,35], license to operate and the guidelines issued by the UK Health and Safety Executive (HSE) provide ample evidence. Experience from other industries suggests the process industries have set standards way above those of other industries.

5.9 Summary

This is a complex road map. From the initial six quantified gaps it is possible to drill down into each of the gaps and identify and apply the many proven techniques for improving manufacturing performance. They cover the whole spectrum of process and manufacturing education and all the techniques have a role and place. The focus is on reducing waste to increase both cost effectiveness and speed of manufacture.

References in Chapter 5

1. Devor, R. and Mills, J.J., 1997, Agile manufacturing research: accomplishments and opportunities, *IEE Transactions*, 29(10): 813–824.
2. Goldman, S., Nagel, R. and Preiss, K., 1995, *Agile Competitors and Virtual Organisations* (Van Nostrand Reinbold, New York, USA).
3. Kidd, P.T., 1995, *Agile Manufacturing* (Addison-Wesley Publishing, ISBN 0 201 63163 6).
4. Heim, J.A. and Compton, P.D., 1992, *Manufacturing Systems: Foundations of World Class Practice* (National Academy, Washington, USA), pp. 100–106.
5. Ahmad, M.M. and Sullivan, W.G., 1991–1999, *Proc of International Conferences on Flexible Automation and Integrated Manufacturing* (Begell House Publishers, New York, USA) http://www.begellhouse.com.
6. Ahmad, M.M. and Sullivan, W.G., (eds), *Int J of Flexible Automation and Integrated Manufacturing* (Begell House Publications, USA, ISSN 1064 6345).
7. Ashayeri, J., Teelen, A.E. and Selen, W., 1995, Computer integrated manufacturing in the chemical industry: theory and practice, *Research Memorandum* (School of Business and Economics, Department of Econometrics, Tilburg University, The Netherlands).
8. Smith, J.B., 1987, System approach for plant reliability, *J of Chemical Engineering Processes*, 83(4): 47–54.
9. Luxhoj, J.T., Riis J.O. and Thorsteinsson, U., 1997, Trends and perspectives in industrial maintenance management, *J of Manufacturing Systems*, 16(6): 437–453.
10. Fraser, A., 1998, Changing the habits of lifetime, *IIR Preventive Maintenance Conference*, 18 February.
11. IACOCCA Institute, 1991, *21st Century Manufacturing Enterprise Strategy* (Lehigh University, Bethlehem, USA).
12. Supply Operations Reference Model (SCOR), http://www.supply-chain.org. Supply Chain Council Inc, 303 Freeport Road, Pittsburg, PA 15215, USA. (Detailed list of references are available on this web site).
13. Benson, R.S., 1997, UKACE lecture: Process control — an exciting future, *Computing and Control*, September.
14. Benson, R.S. and Perkins, J.D., 1996, The future of process control — a UK perspective, *CPC V, 7–12 January, Tahoe City, USA.*
15. Shunta, J.P., 1995, *Achieving World Class Manufacturing* (Prentice Hall, USA, ISBN 0 13 309030 2).
16. Tan, J.S. and Kramer, M.A., 1997, A general framework for preventive maintenance optimization in chemical process operations, *Computers and Chemical Engineering*, 21(12): 1451–1469.
17. Pohlman, M. and Benken, R., 1996, Flexibility with process reliability, *Kunststoffe-Phast Europe*, 86(12):1014–1818.
18. Aldemir, T., Belhadi, M. and Dinea, J., 1996, Process reliability and safety under uncertainties, *Reliability Engineering System Safety*, 52(3): 211–225.
19. Bentley, J., 1999, *Introduction to Reliability and Quality Engineering*, 2nd edn, (Addison Wesley).

20. Schorr, J.E., 1993, *Purchasing in the 21st Century: A Guide to State-of-the-Art Techniques and Strategies* (John Wiley and Sons Inc).
21. McMillan, G.K., 1989, *Continuous Control Techniques — for Distributed Control Systems* (Instrument Society of America, Research Triangle Park, USA).
22. Meredith, J.R. and Hill, M.M., 1987, Justifying new manufacturing systems: a managerial approach, *Sloan Management Review*, 28(4): 49–61.
23. Phall, R., Paterson, C.J. and Probert, D.R., 1998, Technology management in manufacturing business: process and practical assessment, *Technovation*, 18(8/9): 541–553.
24. SERC, 1994, *Innovative Manufacturing — A New Way of Working* (ISBN 1 87066 979 7).
25. Arnold, J.R.T., 1991, *Introduction to Materials Management* (Prentice Hall Inc).
26. Brown, J., Harham, J. and Shivan, J., 1988, *Production Management Systems: A CIM Perspective* (Addison Wesley).
27. Miles, L.D., 1972, Techniques of value, in *Analysis and Engineering* (McGraw-Hill, New York, USA).
28. Pui-Mun Lee and Sullivan, W.G., 1997, Strategic target costing, *Proceedings of the 7th Int Conf on Flexible Automation and Intelligent Manufacturing, University of Teesside, Middlesborough, UK* (ISBN 1 56700 089 4) pp. 851–858.
29. Cooper, R., 1988, The rise of activity-based costing — Part 1: What is an activity-based cost system?, *J of Cost Management*, Summer: 45–54.
30. Tanaka, T., 1997, Target costing at Toyota, *J of Cost Management*, Spring: 4–11.
31. Kaplan, R.S., 1990, Measures of manufacturing excellence: a summary, *J of Cost Management*, Autumn: 44–47.
32. Savage, C., 1996, *Fifth Generation Management* (Butterworth-Heinemann).
33. Kirton, J. and Brooks, E., 1994, *Cells in Industry — Managing Teams for Profit* (McGraw-Hill, UK).
34. *Code of Federal Regulations*, 1998, Chapter 21:
    Parts 110 — FDA Food GMP
    Parts 50, 56 and 312 — Good Clinical Practice
    Parts 211 — FDA Pharmaceutical GMP
    Parts 820 — FDA Medical Advice GMP
    Parts 700–800 — FDA Cosmetics GMP Provisions
    (Government Printing Office, USA).
    http://www.access.gpo.gov/nara/cfr/index.html
    http://www.fda.gov/
35. Department of Health, 1992, *The rules governing medical products in the European Community Part IV, (EC pharmaceutical GMP) Good Laboratory Practice — The United Kingdom Compliance Programme* (HMSO, UK).

Measuring manufacturing excellence practices

6

6.1 Introduction

In benchmarking terms, the important thing is to relate the practices to the performance. It is good practice to group them under similar headings — for example, to have a set of performance metrics concerned with customer service and a set of practices that are believed to deliver these customer services.

This chapter will describe a range of alternative means of describing and measuring the practices. In general, the process industries have been very good at defining and describing the practices, but not so good at measuring the output performance. For example, procedures such as standard manufacturing practice European Quality Model, ISO 9001 and 2, are all means of providing a framework for the practices[1,2]. These will be described in more detail. It is suggested, however, that it is always possible to measure and score the practices in a numeric way. What is more, it is possible to do that for each individual practice and to the sum of all the practices on the physical plant. The net result is it is possible to plot both the performance and the practices on a diagram such as Figure 6.1[3].

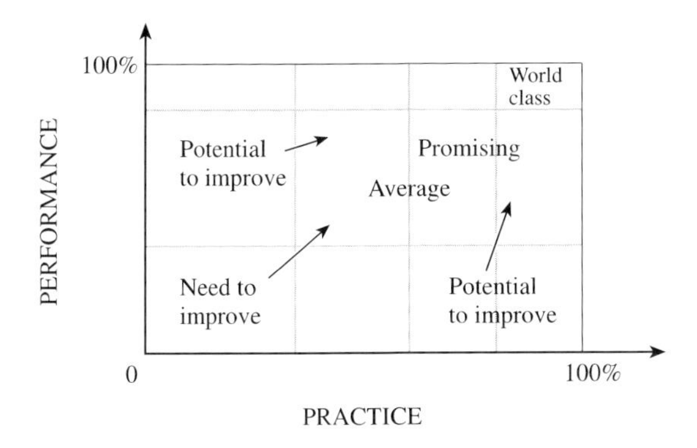

Figure 6.1 Performance and practice benchmarks plotted against world-class manufacturing performance (reproduced by permission of ICI)

Experience in many industries including process confirms that to deliver world-class performance, world-class practices must be adopted and delivered. It is possible to adopt the practices, but not to deliver the performance. These are the circumstances where the plant is using the words but has not taken the practices into the bloodstream of the operation. Such plants have the potential to improve. If a plant has neither the practices nor the performance then it could be described as cither having problems or wonderful opportunities to improve. It is not possible to deliver world-class performance without the practices. This chapter describes how to complete Figure 6.1 for any particular plant.

6.2 Alternative approaches

Over time and in different industries, several frameworks have been derived for both measuring and scoring the process manufacturing practices[4]. This is appropriate, since the practices that matter differ from industry to industry. For example, good manufacturing practice (GMP) in a pharmaceutical industry are particularly appropriate to that industry and would not necessarily be fully appropriate to a large continuous steel plant. Several of the possible frameworks are briefly described.

6.2.1 Good manufacturing practice (GMP)[5,6]

This has been derived primarily from the pharmaceutical industries over many years and is now universally applied in the production of drugs. The key features and headings of the practices are[7]:

- auditing (internal/external/expert);
- calibration and maintenance;
- change control;
- complaint handling;
- contamination control and cleaning;
- contract review;
- corrective action;
- documentation and control;
- hygiene;
- instructions and records;
- labelling;
- management commitment;
- organization and responsibilities;
- policy;
- premises;

- preventative action;
- purchasing and stores;
- quality control and specification;
- statistics;
- traceability and handling;
- training;
- validation (processes, equipment, automation, cleaning and analysis).

A feature of the approach, however, is that an externally-trained facilitator rigorously audits the practices. Aspects such as hygiene and quality are assessed and weighted very highly. Increasingly, it is becoming a requirement of the customers that the plant operates to GMP, and the US Federal Drug Authority is one such external auditor.

6.2.2 The European Quality Model

This model is derived from the Baldridge Award in USA[8,9]. Figure 6.2 illustrates the main components. Total manufacturing performance is scored and includes all aspects. The actual formal assessment is done by an externally-trained assessor and often takes several days intensive discussions in addition to the days of time in preparation. The result is a score between 0 and 1000 and is based on a weighted word model. The weightings are indicated in Figure 6.2 and it has proved very effective, particularly for large successful companies. This approach is, however, very time consuming and is less appropriate for small companies and single plants. Full support for the model is available on a commercial basis[10].

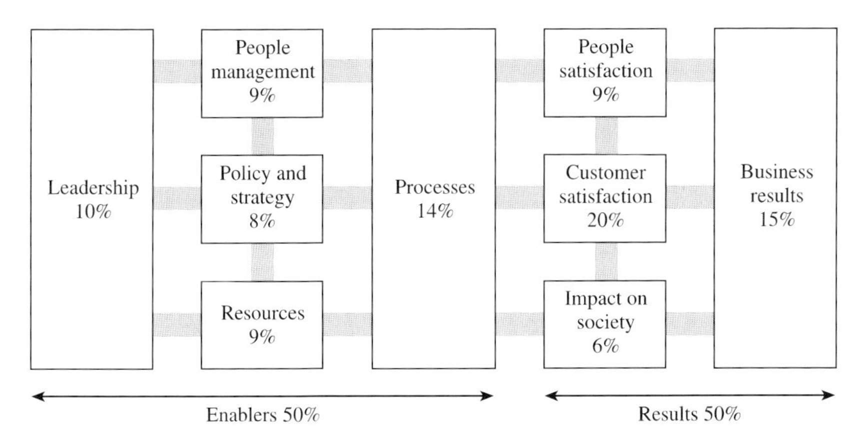

Figure 6.2 EFQM model for business excellence

6.2.3 ISO 9001 and 9002[11–15]

These are procedures which lead to certification at ISO 9001 supplier. The models primarily measure compliances with the firm's stated quality procedures. Figure 6.3 defines the main elements and the scoring is done by an external assessor who uses a word model to validate against each of the headings and hence derive an overall certification. While it would be possible to provide an overall score, this is not currently the normal practice.

6.2.4 MRP2[16]

Over many years the Oliver White Organization has extended the principles of MRP2 from its original remit of materials flow to the whole area of manufacturing excellence. Its guide to operational excellence[16] is a useful document which gives word models to describe all the features of manufacturing excellence. The main headings are summarized below:

- Strategic planning processes;
— qualitative characteristics;
— overview items;
— overview and detail items;
- People/team processes;
— qualitative characteristics;
— overview items;
— overview and detail items;
- Total quality and continuous improvement processes;
— qualitative characteristics;
— overview items;
— overview and detail items;
- New product development;
— qualitative characteristics;
— overview items;
— overview and detail items;
- Planning and control processes;
— qualitative characteristics;
— overview items;
— overview and detail items.

From the word model an experienced assessor, in conjunction with a plant, can derive a score for the overall performance on the particular plant.

ISO 9001: 1994 (formerly BS 5750: Part I)
Quality systems — a model for quality assurance in design/development, production, installation and servicing

- Scope
- Normative reference
- Definitions
- Quality system requirements

— management responsibility
— quality system
— contract review
— design control
— document and data control
— purchasing
— control of customer-supplied product
— product identification and traceability
— process control
— inspection and testing
— control of inspection, measuring and test equipment
— inspection and test status
— control of non-conforming product
— corrective and preventive action
— handling, storage, packaging, preservation and delivery
— control of quality records
— internal quality audits
— training
— servicing
— statistical techniques
— procedures

ISO 9002: 1994 (formerly BS 5750: Part 2)
Quality systems — model for quality assurance in production, installation and servicing

ISO 9003: 1994 (formerly BS 5750: Part 3)
Quality systems — model for quality assurance in final inspection and test

Figure 6.3 Quality system models (ISO 9001, 9002 and 9003)

6.2.5 Manufacturing excellence

Individual companies such as DuPont and ICI have developed their own internal manufacturing excellence approach. The following illustrates the main components:

- manufacturing strategy;
- manufacturing mission;
- SHE excellence;
- operational excellence;
- manufacturing systems engineering;
- financial management;
- maintenance excellence;
- project excellence;
- improvement plans.

These started as a deliberate attempt to capture and describe best practice. This was then used to develop a word model for assessment purposes throughout the organization.

Experience has suggested again that a trained facilitator is best placed to ensure consistency of assessment hence scores. A driver for these in company schemes has to be to reduce the time and effort required in preparation and increase the involvement of the plant teams in ownership of the results. This has proved very effective and again a word model is used to derive a score for the individual topics and overall performance.

6.2.7 The manufacturing excellence diagnostic

Individual consultants such as Dean Kropp and Burke Jackson at the University of Washington have developed and applied their own internal manufacturing excellence approach. The following illustrates the main headings:

People:

- safety, health and environment;
- housekeeping and amenities;
- human resources;
- business philosophy;
- alignment of decisions.

Practices:

- maintenance;
- quality
- management systems;
- supply chain;
- new product development;
- external focus/learning.

Plant:
- capacity and facilities;
- technology;
- sourcing;
- organization.

(reproduced by permission of Dean H. Kropp).

6.2.8 Review

The above are just some of the many approaches that have been used to define good practice in manufacturing. The following summarizes the headings under a number of common business processes:

- manufacturing strategy;
- manufacturing mission;
- SHE excellence;
- operational excellence;
- manufacturing systems;
- maintenance excellence;
- people management;
- project excellence;
- financial management;
- safety/health/environment;
- housekeeping;
- capacity;
- facilities;
- technology;
- sourcing/vendor relations;
- human resources;
- quality;
- production control;
- new product development;
- evaluation systems;
- information management;
- external focus/learning.

This illustrates that while the specific words may differ they all basically adopt a common and similar framework. The first message from this is the framework which should be adopted by a particular plant or business will be driven by the specific characteristics of that business and, in particular, the specific needs of the customers. There will be no single answer for all process industries (see Figure 6.4, page 72).

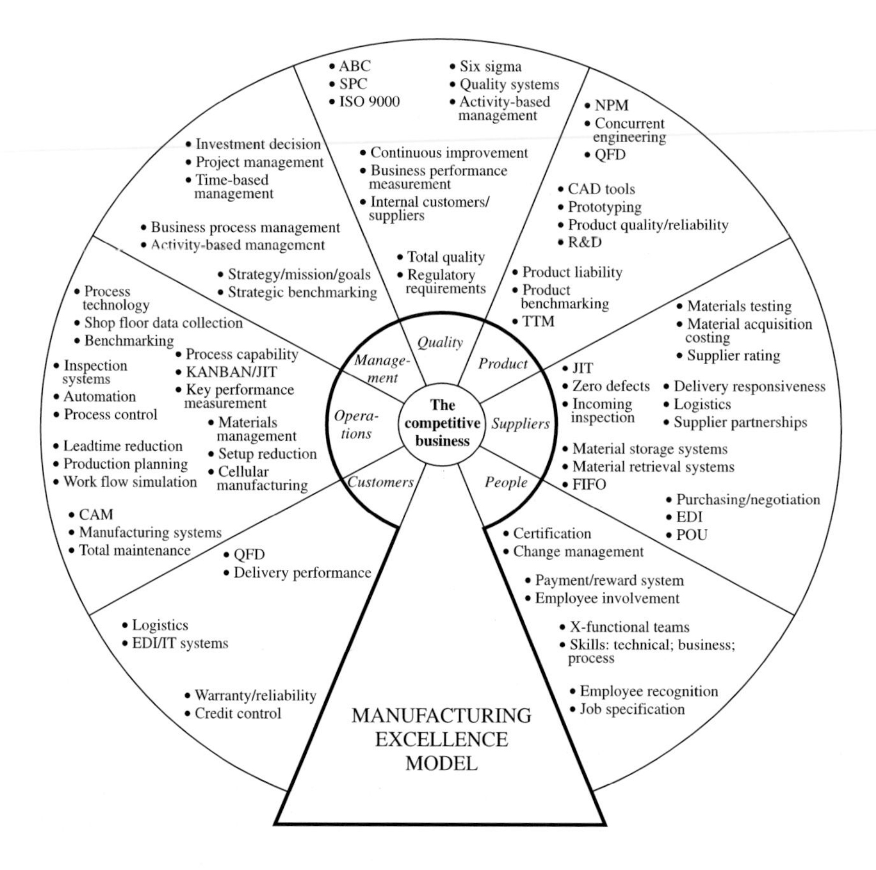

Figure 6.4 Various tools and techniques which can be used for the business excellence model[17] (reproduced by permission of AMT Ireland)

It is also important to ensure that the performance measures selected from those defined in earlier chapters align and are consistent with the practices that are being adopted and measured. It should be possible to identify a direct relationship between input process practices that are measured and the output performance measured results. Figure 6.5 gives a possible framework.

Using this framework, it is possible to plot the score for the performance metric or metrics associated with the input, and the input practices.

While the actual set of practices of priority to a particular business and plant will vary, SHE is common to all the approaches. This is not just because a high standard of SHE is a mandatory requirement to maintain the licence to operate.

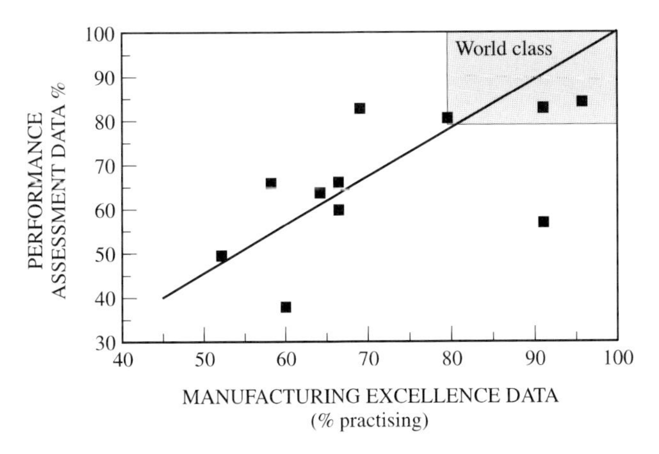

Figure 6.5 Performance assessment versus manufacturing excellence assessment (reproduced by permission of ICI)

A clean and tidy plant where all things are accessible and processes are measured and monitored will not only deliver excellent manufacturing performance but also excellent SHE performance. In fact, some companies use the drive to achieve world-class SHE as the mechanism to focus all the manufacturing excellence activity with great effect.

It is also clear that all the approaches include a significant element on people. Investment in people through training, education, clarity of focus and engagement reap overwhelming benefits when it comes to achieving world-class manufacturing excellence. Therefore, excellence in SHE and people practices are the cornerstone of delivering manufacturing excellence.

The third element of the practice triangle is detailed practices such as operational excellence, maintenance, project execution and new product introduction although none of this happens without clear and consistent leadership, which often manifests itself in a manufacturing mission, and approach to the resulting metrics.

6.3 Scoring the practices

The purpose of benchmarking is to encourage a climate of continuous improvement. It is generally not to generate comparative league tables. Experience suggests that initially when people benchmark they are very reluctant to have the figures displayed or compared. As they gain in confidence, they want to see them displayed in order to see how they compare with others.

For example, UK independent retailers measure all the stores' performances monthly and provide monthly league tables by departments which drives the continuous improvement. If an industry or company wishes to do this, then it is necessary to have some form of scoring system to make it possible. One such scoring system is to measure both the performance and the practices and score these out of 100. Such an approach is used within the Probe benchmarking system[18] and has been used within some of the process industries.

The basic message is that good manufacturing performance derives from good practices which lead to improving business performance. It is possible to be optimistic about the practices but this becomes apparent with poor performance. It is also possible to have some of the performance without all the practices in a country that has a more disciplined approach to such issues. It is, however, impossible to have sustained outstanding business performance without the practices unless there is some form of monopoly.

A means of scoring is to firstly determine the core set of performance measures. For each selected measure, a spread between poor and world class is determined to give each result a score. Figure 6.6 gives an outline suggestion of how this could be done for a set of measures. From this, it is possible to derive a score. In practices, the approach could be different — for example, the plant may be asked if it is in one of the following four categories:

- reactive;
- gaining control;
- commitment;
- aiming for world class.

Against each of these it can give a percentage against a set of questions.

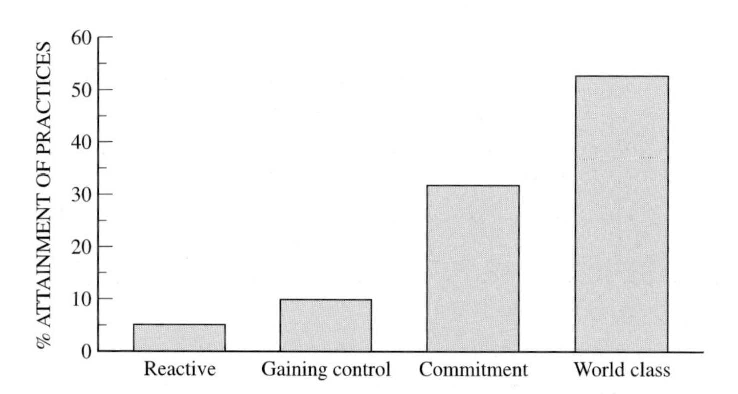

Figure 6.6 Manufacturing excellence practices assessment (reproduced by permission of ICI)

By adding up these scores for every practice, it is possible to derive a percentage figure. The basic message is if scoring and plotting is needed then the first requirement is to have a framework. If an agreed framework is adopted across a plant then it is possible to score and thus benchmark.

6.3.1 Common features of measuring manufacturing practice

All the approaches use a form of word model to score the result. This may be five-step or three-step, but the net effect is a word model. These are ideal since it is important that individual businesses and plants own the word model which they use. Scoring the word model is vital as it gives a measure which can not only be plotted and reported but which can be compared both internally and externally. This comparison is critical to process performance benchmarking and improvement.

Finally, the common feature is a need for an external, experienced and trained assessor. The role of this assessor is to bring an independent objectivity to ensure consistency of interpretation across a number of plants and between a number of industries. This is not to say that the person has to be from outside a particular company but that they need to have the independence and the experience of length of service in the industry and outside to ensure this consistency. They also require facilitation skills as when deriving a score, experience suggests that it should be a mutual decision from the plant or operating team who are responsible for improving the performance. These are the common features of measuring manufacturing practice.

6.4 Summary

It is possible to measure and score manufacturing practices on any process plant. A selection of the inputs is determined by the nature of the industry and the customer base, but it always includes a manufacturing mission, SHE and people. Word models are used to deliver consistency and it is possible to score these on a simple basis. The scoring is important to allow direct comparison with external groups. There must be consistency in relationships between the practices and the performance metrics selected while external independent assessment and facilitation is critical for objectivity. Plotting and measuring the scores regularly and formally reviewing annually is a key requirement to continuous improvement.

Finally, it is never possible to improve all the practices all the time so it is normally better to focus on one or two in each particular time period.

References in Chapter 6

1. Crosby, P.B., 1979, *Quality is Free* (McGraw Hill, USA).
2. Deming, W.E., 1986, *Quality, Productivity and Competitive Position* (MIT Centre for Advanced Engineering Study, Cambridge, Massachusetts, USA).
3. Hanson, P., Voss, C., Blackman, K. and Oak, B., 1994, *A Four Nations Best Practice Study* (IBM Consulting group).
4. Clement, J., Coldrick, A. and Sari, J., 1993, *Manufacturing Data Structures: Building Foundations for Excellence with Bills of Materials and Process Information* (John Wiley and Sons Inc).
5. *Code of Federal Regulations*, 1998, Chapter 21:
   Parts 110 — FDA Food GMP
   Parts 50, 56 and 312 — Good Clinical Practice
   Parts 211 — FDA Pharmaceutical GMP
   Parts 820 — FDA Medical Advice GMP
   Parts 700–800 — FDA Cosmetics GMP Provisions
   (Government Printing Office, USA).
   http://www.access.gpo.gov/nara/cfr/index.html
   http://www.fda.gov/
6. Department of Health, 1992, *The rules governing medical products in the European Community Part IV, (EC pharmaceutical GMP) Good Laboratory Practice — The United Kingdom Compliance Programme* (HMSO, UK).
7. *Management Today's Guide to Britains Best Factories* — Volume 1 (Haymarket Business Publications, UK).
8. Baldridge, M., 1997, *Award Criteria Booklet* (Malcolm Baldridge National Quality Award Office, National Institute of Standards and Technology, MD, USA).
9. Baldridge, M., 1996, *Winners Profile Booklet* (Malcolm Baldridge National Quality Award Office, National Institute of Science and Technology, Gaithersburg, MD, USA).
10. European Federation of Quality Management — EFQM, Brussels Representative Office, Avenue des Pleiadesis, 200 Brussels. http://www.efqm.org..
11. 1994, *BS EN ISO 9001* Quality Systems: specification for design, development, production, installation and servicing.
    1994, *BS EN ISO 9002* Quality Systems: specification for production, installation and servicing.
    1994, *BS EN ISO 9003* Quality Systems: specification for final inspection and test.
    (British Standards Institute, UK).
12. Dobb, F.P., *ISO 9000 Quality Registration* (Butterworth-Heinemann, Oxford, UK, ISBN 0 7506 2532 5).
13. Juran, J.M. and Gryna, F.M., 1993, *Quality Planning and Analysis*, 3rd edn (McGraw Hill, USA).
14. Shaw, J., 1995, *ISO 9000 Made Simple* (Management Books 2000 Ltd, Didcot, UK, ISBN 1 85252 280 1).

15. Prabhu, V.B., Appleby, A., Yarrow, D.J. and Mitchell, E., 1999, The impact of ISO 9000 and TQM on best practice/performance: the UK experience, *4th Int Conf on ISO 9000 and TQM, Hong Kong Baptist University, 7–9 April.*
16. Oliver White, 1993, *The ABCD Checklist for Operational Excellence*, 4th edn (John Wiley and Sons).
17. 1995, AMT Ireland: FARBART (Dublin, Ireland)
18. Matykiewiez, L., 1998, *The Competitiveness Project — Company Benchmarking* (Newcastle Business School, University of Northumbria at Newcastle, UK).

Benchmarking the 'soft' measures

7

7.1 Introduction

The majority of people have experienced visiting a shop, house or company and have formed an immediate conclusion about the operation. There is a distinct buzz and tangible feel about the place. The characteristics which create this perception are very important when forming an overall view of a plant's performance. They are felt and determined as soon as a person 'walks the plant'. This is an integral part of the benchmarking process and must always be completed whether it is performance or practices that are being benchmarked. This chapter provides a framework to measure these 'soft' issues that reflect the plant's culture.

As a way of introduction, consider the Japanese phrase: 'Show me the reception, show me the warehouse and I will tell you the factory'. Now consider two scenarios that could arise when visiting a particular factory.

In the first scenario, you drive up to the factory and park in the visitors' car park which is right next to reception. You walk into reception and are welcomed with a smile and a board stating which company you are visiting. The receptionist has a visitor's tag ready and asks you to sign the reception book. You are then directed to an adjacent area to watch a video or read a card to understand the safety procedures within the factory. The receptionist immediately rings the person who is expecting you. While waiting in a pleasant open sitting area, you observe literature about the company and its annual report, examples of the products are on display plus any prizes and certificates recording success in education, quality, competition, and possibly even some performance metrics of the plant adorn the walls. When the host arrives, you are greeted with a firm handshake, a smile and are taken to start the visit.

Now consider the second scenario. Driving up, you have to hunt for a place to park as the visitors' car park is full. When you arrive in reception, it is empty requiring you to ring a bell or make a telephone call. When you make the telephone call, the person who you have come to see seems surprised. After 10 minutes of waiting, somebody arrives and you are asked to sign a visitors' book in the corner. Whilst waiting, you note that the reception is unattractive with no indication of what products are made and only a picture of the company Chairman on the wall. Eventually, the host arrives in a rather ruffled state, shakes your hand, asks who you are again, and then takes you through to some hastily arranged conference room.

While the above scenarios may seem extreme, they have both happened in reality, creating very different impressions! Now consider the situation when you ask to see the warehouse or the effluent pit. Again, there are two scenarios.

In the first scenario, you walk into the warehouse. The aisles are clear, material is stored on racks and the dates on each rack are relatively recent. Material flows smoothly to the dispatch area and performance measures are on the wall indicating the effectiveness of dispatch and showing the minimal amount of returned material. Similarly, the effluent pit is clean and tidy, the material flows smoothly and there are no obvious signs of disrepair or overspill.

Now consider the opposite scenario where the host is reluctant to show you the warehouse. It is dark on arrival with material piled everywhere. Material stored on the racks looks unsafe, there is no visible labelling on any of the racks and there is a pile of returned material in the corner. You see examples of materials being moved from one rack to another and the floor between the racks and distribution is overcrowded. Similarly, the effluent pit is dirty and there are clear signs that effluent has overflowed into the drains and so on.

These again may seem two very extreme examples but they have both been observed. Already it is clear that the former factory has created a very favourable impression and you anticipate seeing a good manufacturing plant, whereas the second factory created a very poor impression and subsequently you expect the worst. It would take the second factory an enormous amount of presentation and surprising results on the factory floor to dissuade you from the view that this is a poor factory and the practices are not world class.

These examples illustrate what is generally meant by 'soft' measures, but benchmarking requires a more rigorous framework.

7.2 Overall framework

In 1995, Roger Benson was a member of the Innovation Unit at the UK Department of Trade and Industry. Whilst there, he participated in a survey of 120 successful UK companies to try and elucidate the key performance indicators of success. Results of this survey were reported in the *Winning Report*[1]. Within

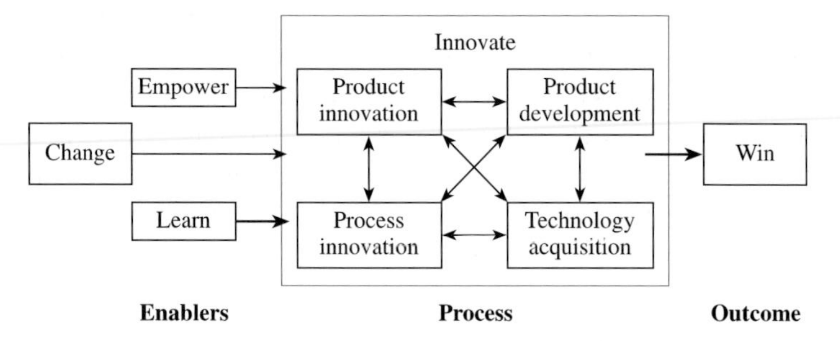

Figure 7.1 Process of winning[1]

this document is a useful framework to assess the 'soft' measures. This is given in Figure 7.1. The framework implies that: *to be a winning company, it is necessary to change, empower, learn, then innovate, to win.*

All the first three things can be achieved but without innovation the company will not be successful. Innovation is defined as the successful exploitation of new ideas. These headings are used to group the 11-element framework that has been evolved by the judges of the UK Best Factory Award[2] judging over six years.

Before moving onto 11 elements, it must be emphasized that this 'walking the plant' happens in conjunction with the performance and practices benchmarking. It is assumed the numbers have already been collected and are available for analysis so this framework is just a focus on the softer issues that support the numbers. For each of the 11 elements, a number of trigger questions are provided which the assessor may wish to use and a number of anecdotes are given to illustrate the type of responses and interpretation.

7.2.1 Change

Is there a cohesive management team that is competent to share the common vision and accept change as the norm?

Invariably, prior to a tour of the factory, the management will make some form of presentation to explain their operations and philosophy. Here, the assessor is looking for the cohesiveness of those present. The following gives a number of possible questions and trigger points for such a management presentation:

- What is your five-year vision?
- If you could have one wish, what would it be?
- What are the keys to your success?
- How do you measure and benchmark your performance?

For example, some companies actually employ people with the job title of 'change manager'. Their task is to ensure there is continuous change and improvement. It is a very clear signal of intent to the management to move in that direction. One would certainly expect to see at this time some form of five or 10-year vision with very demanding targets based on factual or competitive analysis. It is also important to note if the change is a responsibility of one person or whether the whole of the management team shares the vision.

7.2.2 Empowerment

All staff are routinely trained in the necessary skills and the training is measured and displayed. Working conditions are safe, environmentally good and the well-being of staff is clear.

It is a true empowerment of the workforce where improvement groups are active, all results are displayed at all levels and true management exists on the shop floor.

Interestingly, this can be felt immediately when walking onto the shop floor. In an empowered organization, the operatives look you in the eye and are only too enthusiastic to tell you what they do with pride while pointing to measurements on the wall of their improvement and informing you about successes in the market place. The opposite of an empowered organization is a hierarchical organization which is often evident by the feeling of hostility you get the moment you walk onto the factory floor with the manager. Operatives divert their eyes, there are no measurements on the wall and the management try to rush you through.

Some questions to ask operatives at this stage include:

- What does this measurement mean?
- Who is the customer for this product?
- Have you ever met a customer?
- What does this product cost to make?
- Are you happy working here?

Increasingly, companies are displaying their organization as an inverted pyramid as illustrated in Figure 7.2, page 82.

The crucial implication of this diagram is that the operatives are the most important people and are in direct liaison with the customers. If the customers have complaints they come to the operatives and, similarly, it is the operatives who ensure the quality of the product. Such a situation demands that the operative must be fully aware of the true costs of manufacture and be continually educated on the impact of technology and legislation on their job. Therefore, the telling question is to ask the operatives for the true cost of manufacture and

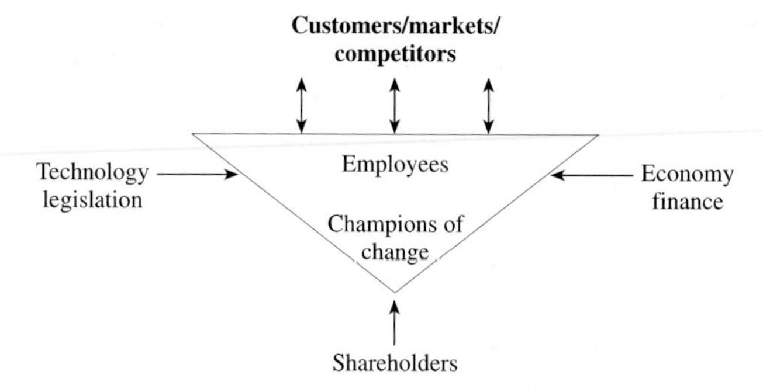

Figure 7.2 Inverting the organizational pyramid[1]

to look for evidence of measures on the wall where true costs are displayed. This is an indication of trust. Training and education is an important part of empowerment. In fact, in world-class companies, training can often amount to an excess of 10 days per year.

Invariably, there will be some form of training matrices visible, with the names of individuals on the rows, and the skills required for a particular job in the columns. In more sophisticated companies, these are measured and given a numerical score. Rates of pay are related to this knowledge. Rooms are available for the training and the philosophy is that management provides the training while it is the operative's job to ensure they become trained.

A useful insight into this whole area is to go and examine the places where the management does not want visitors to go, such as the toilets or the back of panels. If these are untidy and dirty then it is probable that empowerment does not exist.

7.2.3 Learn

Do they know their customers, understand their markets and are they very clear about their unique selling points? Is this understood at all levels of the organization?

Is the manufacturing process responsive to the needs of the market?

In a world-class company, everybody goes to strenuous efforts to learn — to learn from their customers, suppliers, other examples of excellence and competitors wherever possible. Competitors' products and performances are dissected and measured if at all possible. In some cases there is a positive 'pinch with pride' culture which seeks to absorb and adopt learning from all avenues. In this learning culture it is the operatives that go and learn at first-hand, not the management or the marketing people who then pass it on to the operatives. In some cases, the company actually audits the factory for customer readiness. There is a whole procedure covering obvious issues such as the state of reception, measurements, knowledge of the people who show visitors around, display of the products and so on — a very powerful technique for recognizing that the factory is in fact part of the business.

Some questions to test the learning environment are:

- Are you continuing with your education?
- How does your product compare with the competition?
- What is the adherence to the sales and operations plan?
- What is the minimum order time expected by the customer?

The clue to the responsiveness of the factory is to examine the stock, particularly the number of pallets and material on the factory floor and in the warehouse. If the manufacturing process is responsive to needs, then it will be able to make to order and hence not to stock. If it is not responsive then stock will be mounted everywhere.

7.2.4 Innovate

Are they aware of the role of process technology, their competitors' processes and the potential next technology?

Do they actually measure and use the business performance measures, are they widely displayed, correct and understood?

The *Winning Report*[1] states innovation can occur through new products, existing products, the manufacturing process or by acquiring new technology. The implication is that the company must be innovating one of these areas to be successful. It is unlikely that any plant will innovate in all four at the same time. Innovation in the manufacturing process is both a direct benefit to competitiveness and is a route to developing new product features.

Some questions to test innovation are:

- What percentage of today's sales are from products less than four years old?
- How do you define innovation?
- How do you measure innovation?
- What is your competitor's best innovation?

The characteristic of a winning factory is an extensive use of measurement. The key parameters are measured and displayed on the shop floor, available for everyone to see and they are always kept up to date. There is a visible upward trend which in many cases has been exceeded. One of the things that is always measured is product consistency which in itself is a measure of process capability. The use of statistical techniques is just the norm, and if they are not apparent then it is very difficult to see whether the factory can ever achieve its quality or customer aspirations.

It is useful to determine if the managing director is familiar with the competitor's technology, has the director ever been to visit competitors and do they regularly share. In the most enlightened businesses, it is surprising how often competitors visit each other's plants — for example, in the pharmaceuticals industry.

Measurement requires that everybody understands it. Discussion with any of the operatives as to what the measurements mean is usually very informative.

7.2.5 Winning

Winning is a process of growing successfully in the market place. The *Winning Report*[1] noted that to manufacture quality products on time that work is a necessary but not sufficient condition to win in modern markets. In addition, the operation needs a distinctive competence that delivers a competitive advantage. The performance benchmarking described in Chapter 2 has already put numbers against this category. This section looks at the soft issues that illustrate a winning company:

Is the plant a safe and healthy place to work that is concerned with internal and external environmental issues?

BS 5759 or ISO 9000 is the norm, but is it truly in-built into the organization, SPC charts and measurement of process capability?

Are the appropriate systems in place to support the aims of the business?

Is the 'housekeeping' excellent with a place for everything and everything in its place?

Does the material flow logically and simply through the process with the absolute minimum of work in progress ?

We are absolutely convinced that there is a direct relationship between world-class manufacturing performance and safety, health and environmental performance. While it is possible to make a poor manufacturing plant a relatively safe place to work, it is impossible to deliver world-class safety, health and environmental performance without having world-class manufacturing performances.

What is the meaning of world-class safety?

This increasingly is measured as lost-time accidents (LTA) per 100,000 working hours. A good figure would be less than one LTA per 100,000 working hours. The number of working hours is simply obtained by taking the total employees on the site, multiplying that by the hours per week they work and the date and the working days per year. For example, there are many plants in the process industries already exceeding more than 5,000,000 hours which states it is many years since a LTA.

In walking the plant it becomes very clear. If you observe that first-aid cabinets are locked, wash bottles are out of date, temporary cables and tripping hazards on the floor, the close proximity of hot material and no guards on the machines, a feeling that ties need to be tucked in and care must be taken is created. There are no messages on the wall about safety and you begin to be concerned about the safety performance of the plant. Irrespective of what the figures say, there will be a strong feeling that the plant is not safe[3].

Similarly, extraction systems are expected where there are fumes and it is always worth looking at the environmental treatment area to judge the importance of standards set by the factory. This may be the water treatment or an off-gas treatment pit at the back of the factory. We cannot emphasize enough the strong correlation between the SHE performance of a plant and its manufacturing performance.

Are the appropriate systems in place to support the aims of the business?

During benchmarking, the priorities of the business will already have been learnt. They may be responsiveness, consistency or a combination of the two. If, for example, the key priority of the business is to deliver the product within 24 hours of receipt of order with zero defects then it will be disappointing to find that the computer information system is based on a 30-day accountancy period. This suggests that the information system is not appropriate to support the business that is in place. Equally, as has been our experience, to visit a factory where the product is made on the day and shipped the next day and where there are only 15 employees and all of the orders pass through one desk, it may be equally appropriate to have no computer systems at all. The evidence is the more the business runs to a just-in-time operation with zero stocks, the need for such large centralized computer systems such as SAP becomes less and more towards a real-time monitoring and control system. In a world-class factory, the customer is at the centre of the information wheel used not, for example, the accountants or financial function. The key is to ensure that the systems are responsive and align with the needs of the company business.

Does the material flow logically and simply through the process with the absolute minimum of work in progress ?

This again is much more apparent by walking the plant than will ever be learnt in the control room. If, by walking the floor, material is being moved from the front of the factory for the first stage of manufacturing, to the back of the factory for the second stage and then back to the middle of the factory for the third, then you begin to have concerns. Increasingly, a modern world-class manufacturing factory is likely to be a single storey linear operation where material enters at one end and the product is dispatched straight onto the lorries at the far end. Occasionally, this is in the form of a U so that both the goods receipt and product dispatch come from the same place. Unfortunately, some plants are saddled with old assets. This may be a multi-storey factory such as an old cotton mill or one that was built for other purposes. This should not be held against the plant provided they have made strenuous efforts to minimize the movement of material and simplify the whole process. This is the area where it is expected to see simplification at its best. Increasingly, factories that run the KANBAN system, where material is pulled through the factories, operates an excellent MRP2 system.

The best factories will have some form of 'milk-round' system that places the orders on the manufacturing plant and comes back at a prescribed interval to collect the finished goods and order the next material. The speed of flow of material through plants such as this is very impressive.

Is the housekeeping excellent with a place for everything and everything in its place?

This is the final but very important question. A world-class factory is tidy, things are in their place with shadow boards for tools and personal protective equipment. This all becomes possible because material flows rapidly through the factory, off-spec material is not made and thus does not have to be stored, and strenuous efforts have been made to compress time at every occasion. A manifestation of this is a clean and tidy factory.

The opposite is an untidy operation. This is usually a symptom that the plant makes off-spec material which has to be stored in temporary locations for re-processing causing dust to form or there may be spillages and so on. These are all signs of not being a world-class factory. It is amazing how revealing this simple question can be. Often, prior to a benchmarking assessment, strenuous efforts are made to clean the factory before the arrival of the assessors. This does not work. There are always places behind panels, at the back of warehouses, in the toilets and so on where the same effort has not been put in and which will always be very apparent.

This really is where mature judgement comes to play. If the person doing the benchmarking does not have the experience, has not visited many other plants, and cannot give an objective assessment then the process will fail. In addition, one of the tasks of the benchmarking process is to facilitate and encourage the individuals and the plant themselves to start doing the benchmarking themselves and to visit other companies.

7.3 Overall impression

Having visited a plant and completed the benchmarking, the assessor is always left with an impression. This may be described as the 'car park' reflection where the assessor asks the questions. Finally, the following question should be answered: is the company profitable with an excellent future?

7.4 Summary

It is well documented and understood by winning companies that people provide the competitive advantage. It requires the proper systems to be in place and continuous improvement programmes based on change management, empowerment and learning organization culture to be activated.

References in Chapter 7

1. 1995, *Competitiveness — How the Best UK Companies are Winning*, DTI/CBI, June, URN 97/961.
2. Cranfield School of Management, 1999, UK Best Factory Entry Form, *Management Today*, http://www.bestpractice.haynet.com.
3. Benson, R.S., 1997, The link between world class manufacturing and world class safety, health and environment, *Management Today; Manufacturing Excellence*, December.

Review and interpretation of benchmarking

8

8.1 Introduction

The previous seven chapters have introduced the concepts of benchmarking[1,2] and described how to measure the performances, practices and 'soft' measures. In addition, an approach has been presented to quantify the opportunities and use the beginnings of a route map towards prioritizing the next actions. The challenge now is to interpret this mass of information to focus on the key opportunities available for the process manufacturing plant and provide a mature judgement.

This is essentially all about experience and pattern recognition. It is not a simple logical approach of a single number indicating a single factor. In this chapter, through the use of examples, some of the more likely patterns are identified and the suggested priorities discussed.

8.2 Review

All the performance, practices and 'soft' measures discussed in the book may be grouped by business processes since benchmarking focuses on business processes. The experienced reader will realize that interpretation is difficult as more than one business process may affect more than one business measure. For example, a high stock turn may be a result of a problem in the inward and outward elements of the supply chain or may be caused by aspects of manufacturing performance.

To help interpretation, there are a number of tools that can be used to present the data in different ways. These are described in the following sections.

8.2.1 Spider diagram

It is possible to group all the scores for performances on a spider diagram and, if available, to add the scores for the performance (see Figure 8.1).

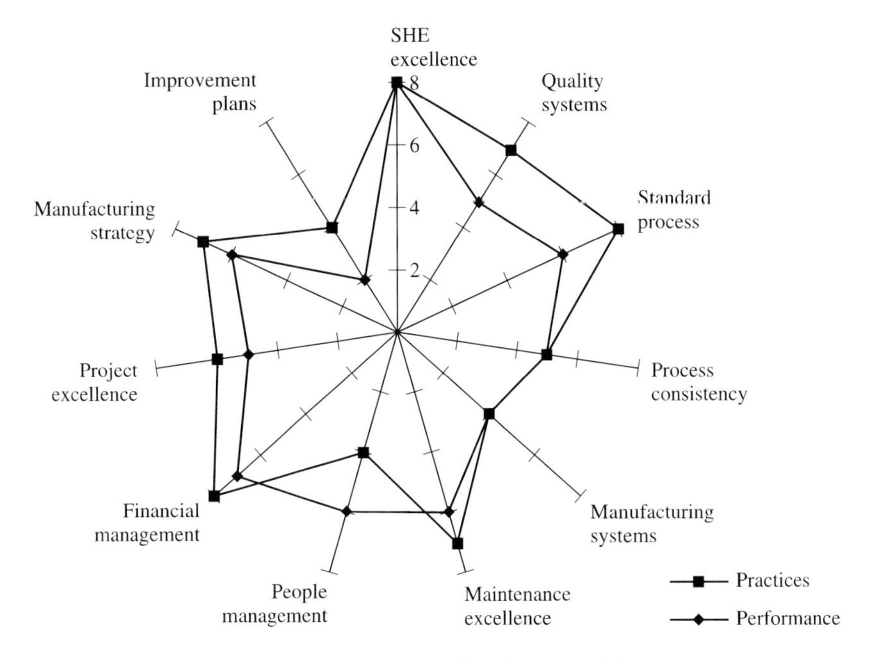

Figure 8.1 Correlation between practices and performance (plant x) (reproduced by permission of ICI)

This approach helps to display the priorities. For example, in the diagram displayed it would suggest that the priority areas are:

- manufacturing systems;
- process consistency;
- improvement plans.

8.2.2 Distributions

This is the approach adopted in the feedback from the UK Best Factory Awards (see Figure 8.2, page 90). It plots a distribution of all the data produced and indicates where a particular factory fits on that diagram. This is particularly useful in the 'soft' areas such as training and absenteeism and so on.

8.2.3 Graphical presentation

Two forms of graphical presentation are useful. The first is the 'box and whiskers' diagram which is illustrated in Figure 8.3, page 90.

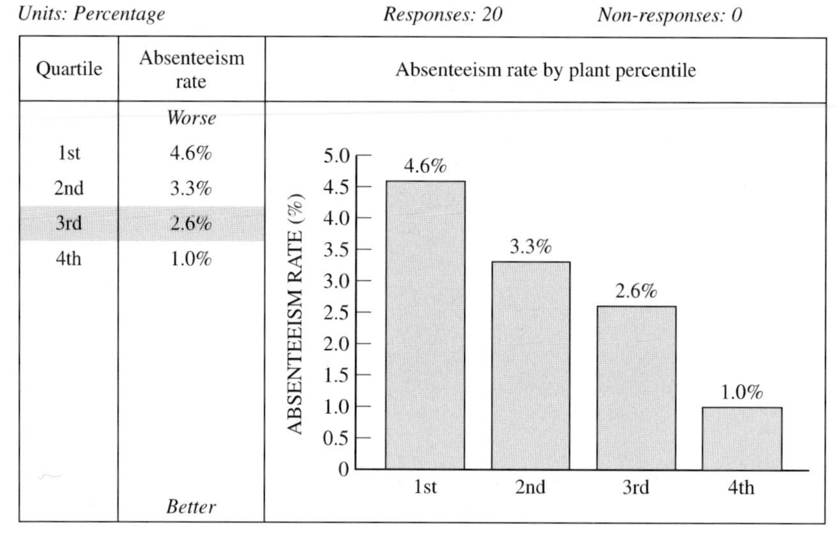

Figure 8.2 Employees: average rate of absenteeism[3] (reproduced by permission of Cranfield School of Management, Cranfield University)

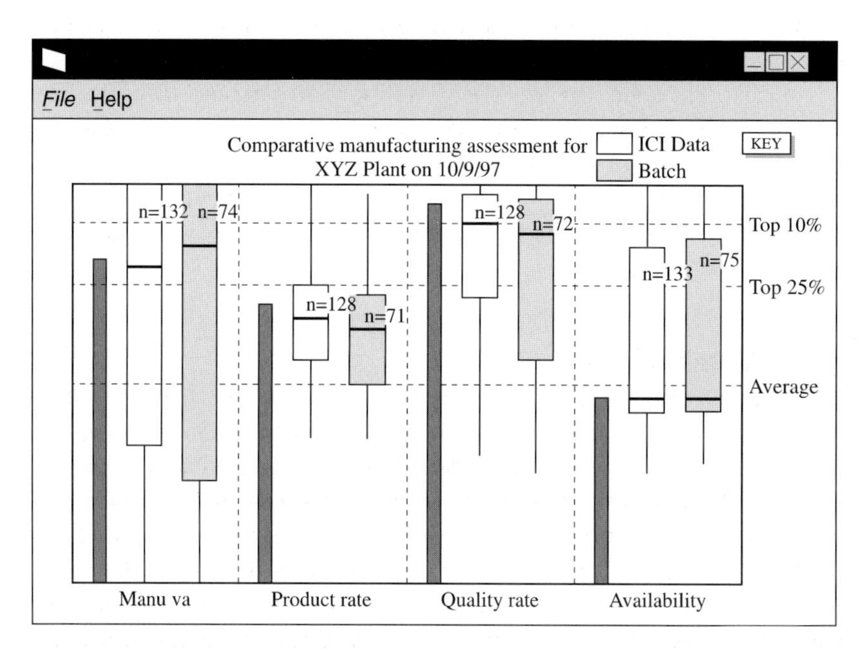

Figure 8.3 Box and whiskers diagram[4]

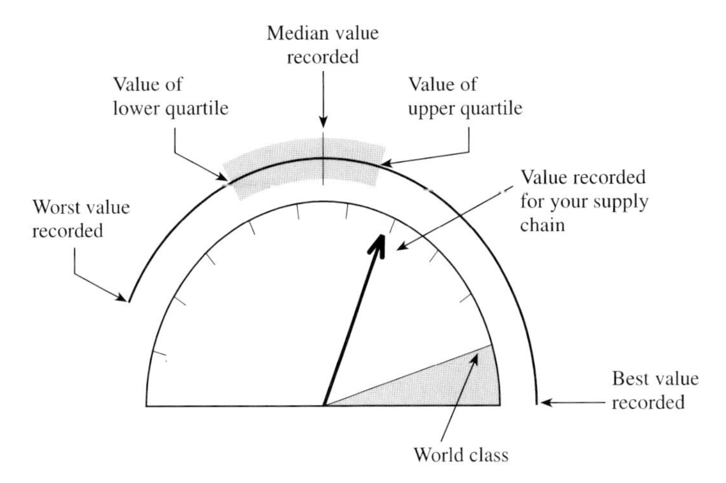

Figure 8.4 Explanation of the dashboard (reproduced by permission of ICI)

The benefit of this is that it communicates the message but not the numbers and hence can be shared among companies without giving away basic information. The principle is the company result is aligned on a score of 0–10, where 10 is world class. Adjacent to that is a box and whiskers where the black line in the middle is the mean for a large set of data, the shaded blocks are the plus and minus 25%, and the black lines give the extreme. By scoring against world-class, all the lines are in the same direction and it is possible to interpret them. This applies to both performance and practices. An alternative presentation of the same information is given in Figure 8.4.

These are all alternative ways of presenting data to communicate both to the person doing the benchmarking and the client receiving the benchmarking message.

8.3 Interpretation

There are no hard and fast rules about interpretation. The analogy is that collecting data is a little like a nurse in hospital taking a temperature. Such data, plus other tests, are given to the doctor and eventually to the consultant. It is the consultant who diagnoses the treatment. In this case, the data collection is the nurse and the task now being faced is to become a consultant to interpret the data. The whole credibility of a benchmarking process hinges on this interpretation. To be effective it must be focused and quite specific with recommended delivery processes. To illustrate it, a number of examples are used in the following section which have been experienced.

8.3.1 Case studies

Example 1 – why is the plant not competitive?

A number of plants are assessed in a particular country. In every case, the measurements, performance and practices in plant reliability are good to excellent. All the measures associated with people are also good to excellent and the SHE performance is average. However, the measures around operational excellence tend to be poor and this is typified by large quantities of stock associated with only average customer delivery performance. The real surprise is that the manufacturing added value per employee is extremely low for a plant of this type. The question is to identify the issue.

The issue turns out to be one where this particular country had for many years been subject to strong tariff barriers which discouraged the import of competitive products. All the plants concerned had been built to satisfy the local market and, while reasonably large, are nothing like world scale. The consequence is that their manufacturing added value was much less than that of a world-scale plant. The basic message is that it takes roughly as many people to run a small plant as a large plant. In recent years, the tariff barriers have been removed and the country is now subject to imports from large world-scale plants situated within 1000 miles. In these circumstances, the costs of import does not exceed the manufacturing economies of these large world-scale plants. Hence, all the plants in this country have a competitiveness issue.

Having made this analysis, the challenge facing the assessor is to communicate it clearly to the business managers who may not realise its full implication. All the plants must be outstanding at reliability, operational excellence, people performance and SHE in order to stand a chance of competing in world markets because the plant scale is not world scale. Hence, the assessor would recommend programmes to move all aspects to achieve world class as opposed to just good performance, in order for the plants to become competitive. This is a classic example of the economies-of-scale argument.

Example 2 – why is the business losing money?

A large continuous plant which has been operating for around 15 years has adopted the principles of manufacturing excellence and this is becoming apparent in the performance. Reliability is increasing significantly, operational excellence measures are improving and the people and SHE performances are more than adequate. That is not to say that there is not scope for improvement, but nonetheless everything is moving in the right direction.

The plant achieves a relatively high manufacturing added value. So what is the problem? The answer is that the business in which this plant is operating is losing money.

The analysis to be drawn from this is that while the plant is making a sufficient margin to be reasonably competitive, the overheads in the rest of the business are pulling it down. Hence, the real priority in this business is to look very closely at the whole overhead structure that is destroying more added value than the plant is creating. If these can be reduced, coupled with further improvement in the manufacturing added value, the plant and the business will then return to profitability.

Example 3 – what is the true OEE?

A food manufacturer supplies a very demanding supermarket and all the customer service messages are outstanding. Equally, the operational excellence measures of quality are very high and the stock turn is very high because of the nature of the products, which are perishable. The manufacturing added value is adequate as well as the SHE performance but there is scope for improvement. The real issue arises in reliability. The first problem is how to measure it. The nature of the production process is that raw materials are received at 5 am, the food is prepared between 7 am and 3 pm and is then rapidly frozen and dispatched to the supermarket by 7 pm. The period between 3 pm and 7 am is filled by a period of maintenance followed by high pressure hosing to keep the line clean and sterile.

The issue to decide is what is the appropriate reliability of such a plant. The plant managed to run from 7 am until 10 pm in the build-up to the Christmas period, hence it is possible to achieve the maintenance between 10 pm and 7 am so that the OEE in the middle of the year would be rather low because the availability was low. The problem to discuss is the fact that the food industry is becoming more capital intensive. As this happens, the focus will move from working capital to fixed capital productivity.

The challenge is first of all to introduce and gain acceptance for the principle of OEE, its measurement thereof and then to introduce the true principles of reliability to improve the asset utilization. This is increasingly happening in the food industry as machinery becomes more sophisticated.

Example 4 – how to improve the stock turn ?

A plant makes a very high-quality low-cost product which is either shipped direct to the customer for pressing and moulding or to a second processor who cuts to size. The performance and practice assessment reports outstanding

SHE performance, good and rapidly improving reliability performance and above average motivation of the employees. The factory has all the appropriate measurements in place and the soft measures are very good indeed. The problem is that the stock turn of the business is very low with virtually no statistical process control and all the customer service measures are below average. This suggests some issue between manufacturing and marketing.

The challenge is to communicate and find a route to explore this issue. It turns out that the manufacturing boundary is very clearly drawn around production, and that supply chain is really treated as a separate issue. OTIF is measured at the dispatch from the factory to the warehouse and not at the final customer. There is in fact a very large and expensive warehouse which, in spite of holding a level of finished stock, does not deliver outstanding customer service. The assessors are aware that there are other companies dealing with sheet products to make floor coverings that manage to produce with no finished goods. Figure 8.5 gives an illustration of how this is achieved[5].

Basically, the order comes in to the factory and is satisfied by cutting to size from a defined stock level. The manufacturing plant only runs to keep the stock level sustained and the warehouse is adjacent to the factory rather than separate. This is a classic KANBAN system[6]. Hence, the recommendation is first to encourage the marketing and manufacturing people to visit an example of this,

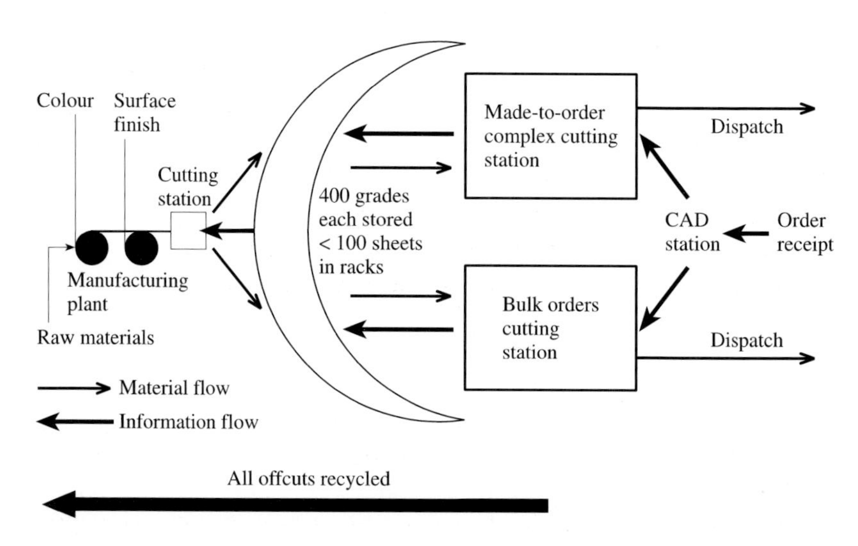

Figure 8.5 Manufacturing process (reproduced by permission of ICI)

then to carry out a supply chain appraisal to quantify exactly the issues with modelling and finally, to move the whole thing forward. A limiting step is the batch size determined by the process equipment. The longer term challenge is to design a minimum batch size of one.

Example 5 – where to start?

A factory operates in several continuous extrusion lines. It scores very poorly on customer service levels, reliability, operational excellence and people motivation. This is supported by a walk of the factory where the scores on all the soft measures are also low. It is untidy and disorganized and the warehouse is both huge and disorganized.

The challenge is where to start. Even an inexperienced eye would observe that there is scope for improvement, but it is a massive challenge. In these circumstances, the normal recommendation is to start with the people — to work extensively with the operators to develop a culture of continuous improvement and to put in the first steps of the visible factory. This factory is moving on a long journey to continuous improvement which could take five years or more.

8.4 Summary

These are just some of the interpretations that can be developed from process industry benchmarking[7–11]. They illustrate the importance of experience which is only gained through time — that is, both time in industry and time in assessing and benchmarking a number of plants. It is one of the many reasons why it is always beneficial for two people to benchmark, so that there are two pairs of eyes and ears. The provision of a benchmarking database and absolute measures is critical in gaining credibility and acceptance of the figures. It must always be possible to relate these back to specific sources. In addition, the access to a compendium of case studies and the ability to arrange visits to see and feel excellence in the area focused upon is another very important supporting mechanism in the reviewing and interpretation of results[12].

References in Chapter 8

1. Phall, R., Paterson, C.J. and Probert, D.R., 1998, Technology management in manufacturing business: process and practical assessment, *Technovation*, 18(8/9): 541–553.
2. Sheth, J. and Eshgi, G., 1994, *Global Strategic Management Perspectives* (SW Publishing Company).

3. New, C., 1998, *UK Best Factory Award Benchmark Report* (Cranfield University, UK).
4. Benson, R.S., 1997, Benchmarking lessons in the process industries, *Management Today: Manufacturing Excellence*, May (Haymarket Business Publications, UK).
5. Landvater, D.V., 1993, *World Class Production and Inventory Management* (John Wiley and Sons Inc).
6. Wallace, T.F., 1993, *Customer Driven Strategy: Winning through Operational Excellence* (John Wiley and Sons, UK).
7. De Toni, A. and Tonchia, S., 1996, Lean organisation, management by process and performance measurement, *Int J of Operations and Production Management*, 16(2): 221–236.
8. Fitzgerald P., 1996, Benchmarking pays off, *Chemical Marketing Reporter*, 249(15): SR16–SR17, April 8.
9. Meredith, J.R. and Hill, M.M., 1987, Justifying new manufacturing systems: a managerial approach, *Sloan Management Review*, 28(4): 49–61.
10. Mullin, R., 1996, CMA's satisfaction index, *Chemicals Week*, 158(41): 45.
11. 1992, Future organisation model, *Business Week*, Nov 30: 63.
12. Kay, J., 1993, *Foundations of Corporate Success: How Business Strategies Add Value* (Oxford University Press, UK).

The need for agile manufacturing

9

9.1 Introduction

Benchmarking process manufacturing is relatively new to the process industries, which is why this book has been prepared. It moves the focus from process efficiency to manufacturing efficiency. The evidence from the assessment of many process plants is that while there are pockets of outstanding manufacturing performance, the overall performance of the process manufacturing industries is average. Hence, the potential for improvement from the existing assets is very large. It is, however, an improvement journey where the appropriate processes or techniques are determined by the present position. Figure 9.1 provides one description of such a journey. An alternative view is provided by the journey to excellence in Figure 9.2, page 98.

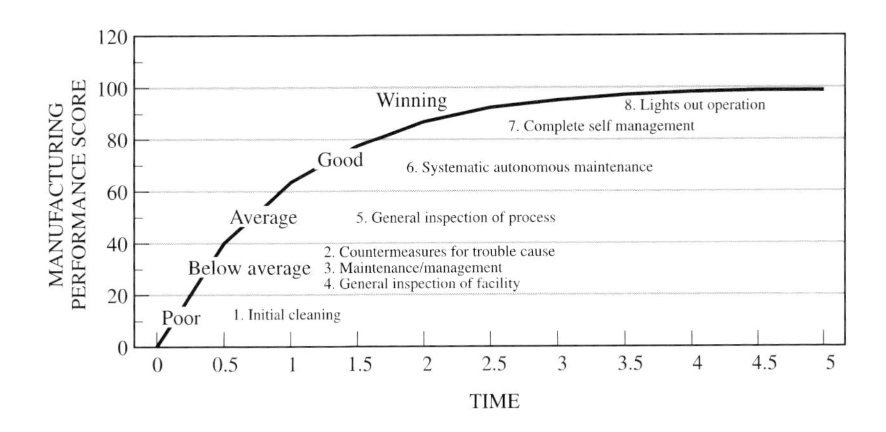

Figure 9.1 The manufacturing improvement journey through TPM

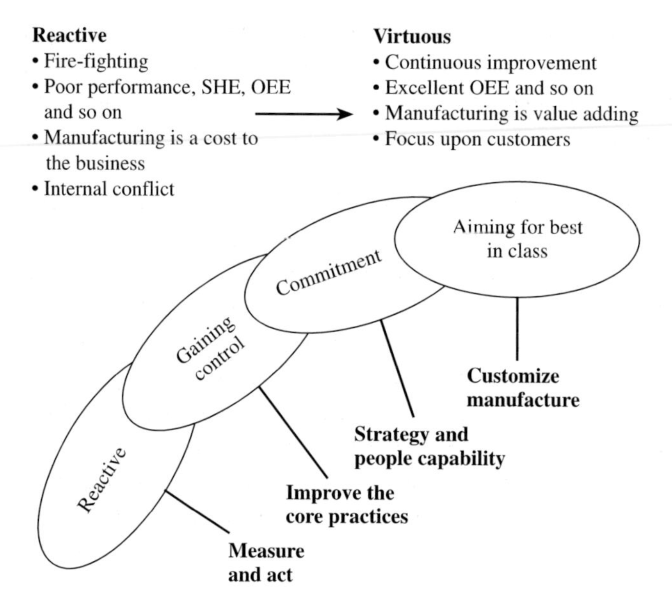

Figure 9.2 The journey to world-class manufacturing

Both diagrams illustrate that improvement is a journey which takes years of focused and consistent attention. As world class approaches, the challenges become more technical while opportunities become clearer.

One characteristic is that when an operation approaches world-class manufacturing performance, it will become closer to the dynamics of the final customer[1]. This always increases the need for agility. This chapter explores these issues beginning with an historical perspective.

9.2 Historical review

Process manufacturing is a little like the curate's egg; it is good in parts. This is illustrated by considering two similar hydrocarbon processes[2].

Firstly, a large hydrocarbon process manufactures a precursor to nylon. It consists of four large reaction vessels in series. Hot hydrocarbon passes through the vessels where it mixes with air in the presence of a catalyst. The conversion level is small and so the hydrocarbon is recycled several times. The plant is large and would cost millions of pounds to build today. It is controlled by a large central control room which, until recently, had a conventional panel control scheme. In addition, there is a high integrity protector system (HIPS) which is used to ensure safe operation.

When we first experienced this process in 1970, it would typically have a process yield of around 80%, the plant required an annual shutdown of 10 days for maintenance and there were two cleaning shutdowns. A team of eight operators per shift worked on a four-shift system, supported by two instrument artificers, two mechanical technicians and an electrical technician. The plant looks very similar 30 years later and is still operating, the panel control system has recently been replaced by DCS, the instrument and electrical technicians have combined, the yield has increased by around 5% and the frequency of shutdowns has extended to two years. The philosophy of control has changed very little, other than where it has been possible to install more analysers, and the HIP systems are still operating reliably. This is not an untypical story of a hydrocarbon process.

There is, however, an alternative represented by another hydrocarbon processing plant. This also consists of four hydrocarbon and air reactors but in this case they operate in parallel producing chemical gases and power. This plant was very unreliable 30 years ago, it was difficult to run in cold weather and start-ups could be a problem; the operator tended also to be the maintenance person and sets of tools were always visible. Control was fairly rudimentary with a few mechanical and pneumatic controllers, and efficiency was in the low 30%s. Today, things have moved on dramatically. The efficiency has doubled, the period of shutdowns has gone out by a factor of three, the reliability is extremely high and the start-ups are no longer a problem. In fact, operating the plant has become so boring that more attention is focused on the creature comforts of the operators such as what type of music system to use.

By now, it might be easy to guess that the second process is in fact a motor car. Over a similar 30 year period, the performance of the motor car has improved dramatically and process control has been a major component. Such is the degree of sophistication of control within a car that the average person is no longer able to maintain it, and it requires sophisticated computer monitoring to control it. The net effect is that the car is much more reliable than it ever was before and, in addition, its price has fallen.

This contrasting performance illustrates the current issues in process manufacturing. In the process industries, there are some world-leading and some world-lagging process manufacturing performances.

Whilst some manufacturers exploit the best process research in the world using all the latest tools, there is, however, a large tail of average to even poor process manufacturing and one is left wondering why this is so.

9.3 Process manufacturing competitiveness

This book suggests advances in process manufacturing occur where the industry is under strong commercial and technical competitive pressures[3]. Cars and aircraft are classic examples but, similarly, the steel industry has also undergone such pressures. By contrast, industries that are protected by Government, raw material supply restrictions or patents have not been under such manufacturing pressure and thus have not exploited process control.

Manufacturing in this sense means all aspects from the raw material through to the final finished product. It is much more than simply the chemical process and includes all aspects of purchasing and distribution. For these reasons, process manufacturing efficiency has been measured in exactly the same way as any other manufacturing industry, as summarized in Table 9.1.

Table 9.1 Measurement of competitiveness for the process manufacturing industries (reproduced by permission of ICI)

	Poor	Good	World class	
OTIF %	40%	99.9%	> 99%	Determined by the market
Customer complaints (% of orders)	6%	0.01%	< 0.1%	
Adherence to production schedule (%)	40%	99%	> 95%	
Production rate	60%	99%	> 90%	Determined by the process, strongly influenced by control
Quality rate	45%	99%	> 99%	
Availability (%)	70%	96%	> 95%	
OEE	20%	94%	> 85%	
Process capability (*Cpk*)	0.6	1.5	> 2	
Stock turn	4	19	> 25	
Absenteeism	10%	0.8%	> 1%	Culture and management style

The table is built upon a set of measures driven by the customers on questions of delivery and quality, a set of measures on the actual performance of the manufacturing assets and finally, a measure of motivation of the employees. Excellent, reliable operation of the assets leads to the production of quality products on time and in full which delights the customers and normally ensures highly-motivated operating personnel.

A means of measuring the effectiveness of the operating assets is to adopt the measure of overall equipment effectiveness as described earlier in the book. An OEE of 100% would be achieved by a process plant that always operated at its maximum proven rate; where all the product produced was perfect and where the plant never shuts down. A very demanding target indeed, but world class in the process industries is recognized as having an OEE in excess of 95%. It is present in those examples where world-class process control is observed. Typical performance in the UK process industries for OEE will vary from 50% to 70%. Given that many of the companies operating at these lower OEEs are still profitable would suggest that they are not yet subject to the competitive pressures of some other industries.

Competitive process manufacturers are demanding the production of highly-consistent products from processes offering agility of manufacture thus enabling them to deliver on time in full when the customer demands it. The requirement is process innovation.

9.4 Innovation in the process industries

Innovation is defined as the successful exploitation of new ideas. These new ideas apply to all aspects of the process industries and manufacturing from raw material through to finished goods.

The direction of the market is very clear in demanding more customer focus, bespoke solutions, and the rapid development of new products, processes and response to change. This is often called agility.

This is set against a background of rapidly changing consumers who are much more IT-literate through the advancement of computer software and the existence of the Internet, and there will continue to be an oversupply in the process manufacturing market. Pressures are on to remove patent, Government and all other aspects of protection. This is a topic currently being addressed by the manufacturing panel of the UK *Foresight* exercise[4,5]. This sees a future process manufacturing with three characteristics as illustrated in the following sections.

9.4.1 Bulk manufacture

The bulk industries represent process manufacturing as it is perceived today — that is, large plants on large sites operated professionally and safely by well-trained teams of specialists. The drive in the future will be the optimization of these assets. Aspects such as the capital, operation and distribution of the products will be included. In fact, all the tools probably already exist today and advances in computing mean that the hardware to achieve this is possible[6] (see Figure 9.3).

While these large plants will continue to exist, they have the limitation of not being very agile in rates, products and quality.

9.4.2 Agile manufacture – intermediate

To seek agility[7–11], the industry is already moving to various types of intermediate manufacture. The characteristic of this is that the plants are smaller and tend to be nearer the customer. The key drivers to this are customization of size, quality, service and effect. One of the longest examples of such has been in the air liquefaction industry. All these process plants are trying to achieve agility as illustrated in Figure 9.4.

Historically, low-cost manufacture has been achieved by building large single-stream plants. This has, however, resulted in the lowest cost at the exit of the plant, and not necessarily at the customer's premises. Where flexibility is required, the conventional route has been to build a batch plant which, within bounds, could make almost any product needed. What is now being sought are plants which have the economies of a large single-stream plant with the flexibility of a batch plant. Two responses to this are apparent.

Today's industry is perceived as a bulk chemical manufacture

Present	*Future*
• Oil	• CAD/CAM
• Olefins	• Advanced control
• Base white paint	• Low wage economies
• Methanol	
• MMA	
• MDI	
• PTA	

Driven by optimization of capital, operation and distribution costs

Figure 9.3 Comparison of the present and future of bulk chemical manufacture (reproduced by permission of ICI)

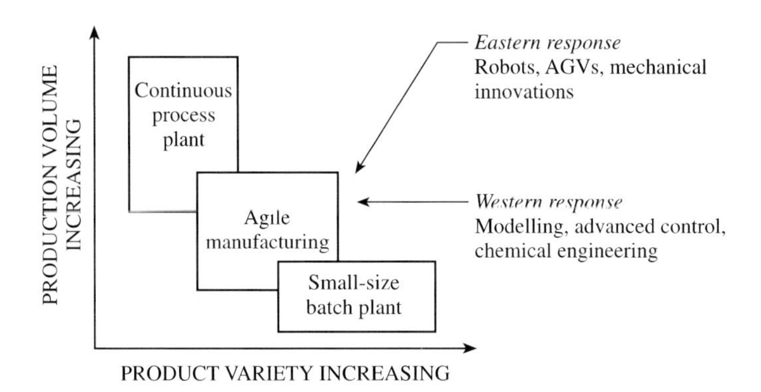

Figure 9.4 Responses for agile manufacturing (reproduced by permission of ICI)

The first, which we refer to as the 'Eastern response', is based on exploiting the technology of automated guided vehicles, positioning sensors and robots. A typical process plant in this area is a pipeless plant which is where the vessels containing the chemicals are moved between filling stations, processing stations and product filling stations on automated guided vehicles. Such plants already exist both in the East and West and are setting new standards in flexibility and quality. Another example is plants on the back of a wagon where the chemical manufacture and mixing only occurs at final delivery as is currently used for the distribution of explosives.

The 'Western response' builds on the strength in the West for process modelling and control. The principle is that if a continuous plant is taken, a perfect dynamic model developed, then it is possible from the model to predict what changes to make to the process control set points to produce different products exactly on demand. Such plants and techniques are now being exploited.

A characteristic of this move to intermediate manufacture is the growing dependence on process control. Again, however, the tools and techniques already exist and this is a question of exploitation, knowledge and culture to deliver the improvements[6] (see Figure 9.5, page 104).

9.4.3 Consumer manufacture

It is suggested that the future for some process manufacturing lies in manufacturing the chemicals at home with suggestions such as making methanol in the garage, paint at the supermarket, treating dirty water in the house, and pulping recycled paper in the refuse vehicle[12]. There are already examples of outdoor swimming pools that are slightly salted and the chlorine is manufactured electrolytically in the filtration system, small bedside oxygen plants for invalids and so on. A classic example would be the house's central heating boiler which is a form of self-contained consumer-based chemical plant![6] (see Figure 9.6).

Interestingly, much of the research for these plants was developed in the UK in the 1980s under headings such as intensification, but had trouble finding commercial outlets because the philosophy was still to build large plants based

To intermediate chemical manufacture closer to the customers

Present	*Future*
• Effluent	• Pipeless plants
• Industrial chemicals	• Plants on wagons
• Mixtures	• In-store and company plants
• Energy	• C1 – > C2 chemistry
• Films	• Alliances with equipment manufacturers
• Adhesives	

Driven by customization — size, quality, service, effect
Agility in all aspects

Figure 9.5 Present and future of chemical manufacture (reproduced by permission of ICI)

This will lead to consumer chemical manufacture

Present	*Future*
• Saltwater pools	• Throw away plants
• Bedside oxygen	• Dirty water treatment
• Greenhouse gases	• Flue gas cleaning
• Central heating	• Water-based paint
• Beer and wine	• Heat pumps
• MDI foam	• Methanol from methane
	• Explosive growth

Driven by low cost high volume manufactured technology using UK research from the 1980s

Figure 9.6 Consumer focused chemical manufacture (reproduced by permission of ICI)

Process engineering

- Operating in the laminar flow regime
- Spinning processes — 'High G'
- Membrane technology
- In-pump mixing of viscous fluids
- Jet mixing
- Mono particle drying
- Bioprocessing
- Electrolysis
- Nano technology
- Biological catalysts

Process control

- Fuzzy control
- Neural networks
- Optimal control
- Plant-wide management
- Supply chain modelling
- Stochastic control
- Smart sensors
- Analysers on a chip

Figure 9.7 Examples of existing UK research offering potential for future chemical plants (reproduced by permission of ICI)

on the economies-of-scale philosophy. Some of the topics are illustrated in Figure 9.7[4].

This also illustrates that many of the process capability techniques exist, but they also had difficulty finding exploitation in the process industries because the market was not and still is only rarely demanding such properties. This contrasts with electronics, cars and food where criteria such as six sigma performance are commonplace. Hence, as the process manufacture moves towards consumer manufacture, the need for improved process capability will advance and the plants will become as critical on operation as the existing motor cars.

The key element in making this change will be significant reductions in the price of the process operating units (see Figure 9.8, page 106).

As the manufacturing cost of the unit goes down, the margin per tonne may go up although the degree of technical sophistication may not be as high as large plants. The good news is that this is potentially a very large market for industry. The area of concern, however, is that some countries' histories have come from large manufacturing plants therefore they may be more resistant to change compared with other industries with no such history. Readers will recognize that this is a very similar story to the move from large mainframe computers to the PC with all the apparent market growth and change.

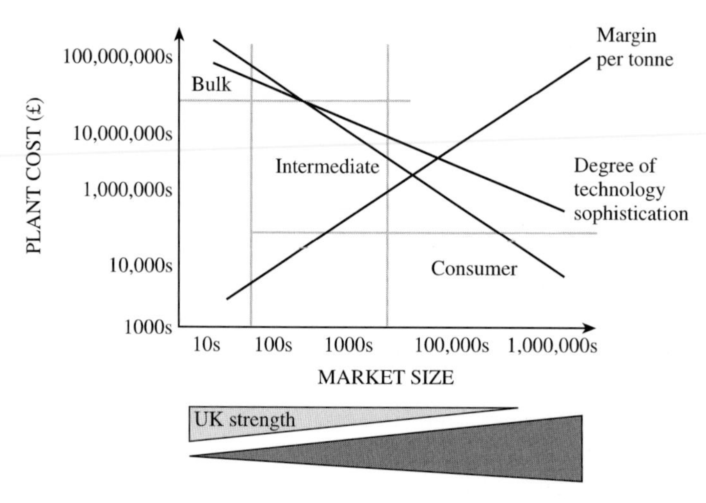

Figure 9.8 Potential impact of reduced plant costs (reproduced by permission of ICI)

9.5 Challenges

To meet this future vision, agile plants will need to deliver the following winning challenge for the process industries of the 21st century:

'Process plants are achieving overall equipment efficiencies (OEE) in excess of 95% with operators who are regularly in contact with the customer, and where no alarms are visible and no safety, health or environmental issues arise.'

Meeting this will require technological, cultural and structural challenges (see Figure 9.9). Culturally, the desire to learn from others, to share and to work in multi-functional teams will continue to increase. There are already some encouraging signs in this area.

9.5.1 Results in process manufacturing

The result of all this change will be an industry that moves from one that exploits data through to one which uses the data to produce wisdom (see Figure 9.10).

In this move, the added value will come from the provision of knowledge and expertise as embedded in software, rather than the provision of hardware and plant. Concepts such as smart unit operations can be envisaged which, when plugged together, identify their requirements and establish themselves automatically. This is similar to the loading of software onto a computer at the present time. It is inherent in the proposals that process design and control

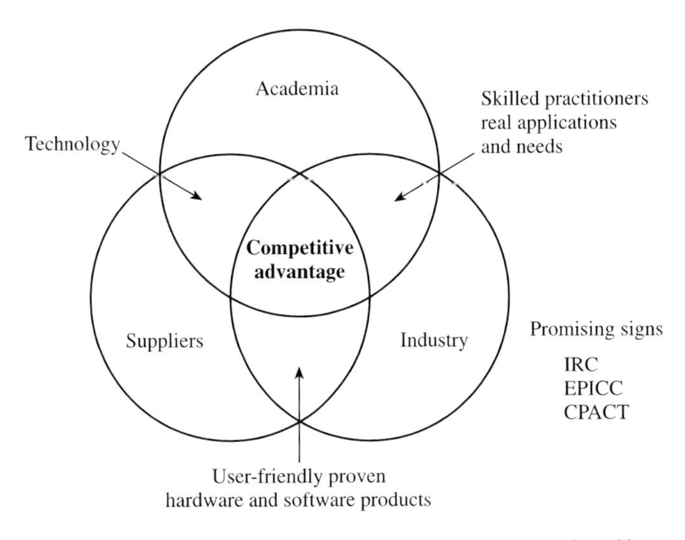

Figure 9.9 Academia/supplier/industry partnership (reproduced by permission of ICI)

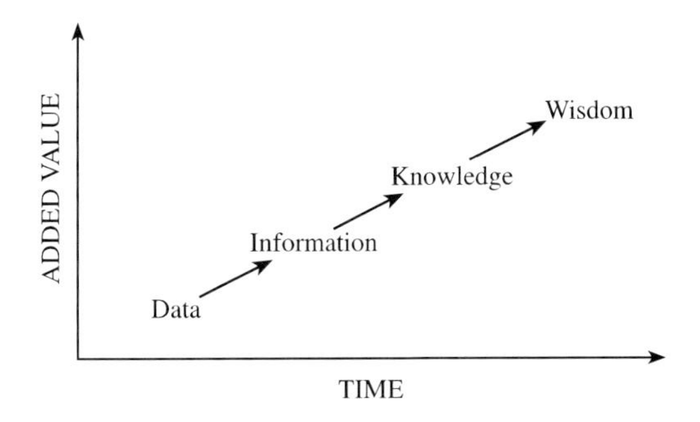

Figure 9.10 Moving from data to wisdom (reproduced by permission of ICI)

must be totally integrated with no functional barriers whatsoever and that this design will be totally dynamic. The resulting processes for improving manufacturing performance are:

- ‘Smart’ unit operations:

— self commissioning, repairing and safe
— no alarms with very few trips
— simple to understand and use advanced process control;

- Designed as an integrated system[13]:
— process and control
— dynamic total manufacturing models;
- Integrated by 'smart' optimizers:
— dynamic
— global.

In fact, this book suggests that in the future, processes will be driven by the need to improve manufacturing competitiveness as measured by process benchmarking.

9.6 Summary

Process manufacturing benchmarking demands a change of culture and thinking from one of large-scale single-stream continuous and batch plants to small consumer plants based in individual homes. The market size potential is enormous, but speed will be essential if this is to be delivered[14]. Process benchmarking is the key tool.

References in Chapter 9

1. Benson, R.S., 1997, UKACE lecture: Process control — an exciting future, *Computing and Control*, September.
2. Benson, R.S., 1989, Process systems engineering: past, present and a personal view of the future, *Computers in Chemical Engineering*, 13(11/12).
3. SERC, 1994, *Innovative Manufacturing — A New Way of Working* (ISBN 1 87066 979 7).
4. Office of Science and Technology, 1995, Progress through partnership: manufacturing, production and business processes, *Technology Foresight* (HMSO, UK, ISBN 0 11430 123 9).
5. Office of Science and Technology, 1995, Progress through partnership: chemicals, *Technology Foresight* (HMSO, UK, ISBN 0 11430 117 4).
6. Office of Science and Technology, 1998, Processing the future, *Report of the Foresight Process Industry Group* (DTI, UK).
7. Devor, R. and Mills, J.J., 1997, Agile manufacturing research: accomplishments and opportunities, *IEE Transactions*, 29(10): 813–824.
8. Gardiner, K.M., 1998, Globalization, integration, fractal systems and dichotomies, *Proc of 8th Int Conf on Flexible Automation and Intelligent Manufacturing* (Begell House Inc, New York, ISBN 1 56700 118 1), pp. 15–27.
9. Goldman, S., Nagel, R. and Preiss, K., 1995, *Agile Competitors and Virtual Organisations* (Van Nostrand Reinhold, New York, USA).
10. Kidd, P.T., 1995, *Agile Manufacturing* (Addison-Wesley Publishing, USA, ISBN 0 201 63163 6.

11. Wiendahl, H.P., 1994, Management and control of complexity in manufacturing, *Annals of CIRP*, 13(2): 533.
12. Benson, R.S. and Ponton, J.W., 1993, Process miniaturisation — the route to total environment acceptability, *Trans IChemE, Part A, Chem Eng Res Des*, 71(A2): 160–168.
13. Crooks, C.A. and Macchietto, S., 1992, A combined MILP and logic-based approach to the synthesis of operating procedures for batch plants, *Chem Eng Comm*, 114: 117–144.
14. Welch, J., 1993, Lessons for success, *Fortune*, Feb 25: 8.

10 Potential impact of process engineering on world-class benchmarks

10.1 Introduction

The earlier chapters of this book provided a framework for benchmarking an existing manufacturing process. They defined how to measure the performance, practices and culture, and suggested some existing world-class benchmarks for comparison. The majority of these world-class benchmarks have been derived from non-process industries and confirmed by experience from the process industries — for example, customer service from the retail industry and reliability from the aircraft industry. A characteristic of these industries is that they serve the most demanding customers.

Process engineering has, however, a long and successful record of innovation in response to market pressures — for example, the low pressure polythene plant, development of new polymers and advances in biotechnology. There is also strong evidence of the world process industries becoming far more competitive as their customers become more demanding of them[1–14].

This final chapter suggests some areas where process engineering, using existing knowledge and future research, could and probably will move the world-class benchmarks for the process industries. It concludes with a set of manufacturing benchmarks for the process industries in five years time. Much of the contents are speculative and are provided as a stimulus to future thought, rather than a statement of the present situation.

10.2 Framework

As a reminder, the framework which has been adopted in this book for benchmarking is as follows:

'A world-class process manufacturing plant delivers outstanding safety, health and environmental performance, exceeds customer requirements from very reliable assets, exhibits operational excellence and is operated by highly-motivated people'.

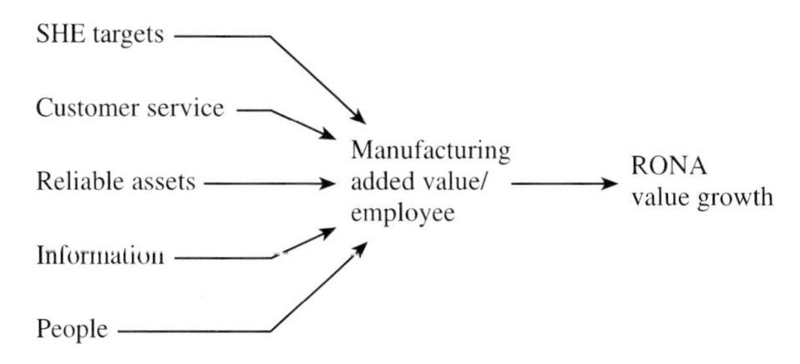

Figure 10.1 Assessing manufacturing performance (reproduced by permission of ICI)

This is summarized in Figure 10.1. Each element is examined in detail, though a number of common features become apparent.

10.2.1 SHE

This is the one area where the process industries already set the world-class benchmarks. It is widely recognized that other industries look to the likes of DuPont and ICI for the world-class standard. The 1994 safety performance of many of the leading players in the industry is summarized in Figure 10.2, page 112. While there is a wide variation in the performance achieved today, all these companies achieve standards well in advance of those of most other manufacturing industries. In this area, it is relatively easy to define what is the eventual world-class standard. Zero SHE incidents will be the future world-class benchmark.

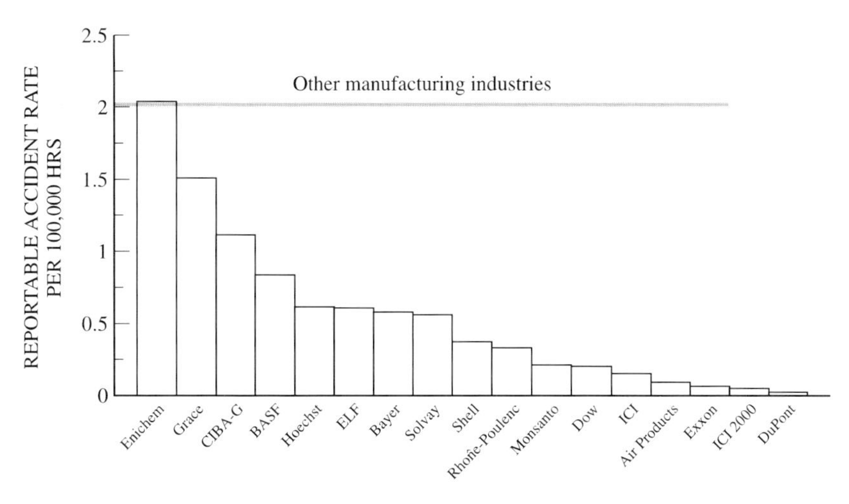

Figure 10.2 Accident rates comparison 1994 (reproduced by permission of ICI)

As Figure 10.2 illustrates, many companies are striving very hard to achieve that standard within the constraints of the present process engineering. There are, however, a number of directions that process engineering could take to achieve the target of zero. These include:

- Reducing the plant volumes to almost zero:

— intensification, plants on a micro chip and so on;

- Moving to 'lights out' operation:

— fully automatic start-up, shutdown and fail-safe

— intelligent equipment failure measurement

— multivariate statistical process control

— small distributed plants;

- No waste products:

— recover and use

— processed to carbohydrates

— never made due to precise control and measurement;

- 'Poka Yoke' design[15,16].

Some of these directions are described in the following sections.

10.2.1.1 'Poka Yoke' design

Many in the process industries may not be familiar with the concept of 'Poka Yoke'. In Japanese, this means fool proofing.

The techniques have been developed over many years in the electronics and automotive industries. It is basically an approach of design where it is virtually impossible to make a mistake. In the non-process industries, it takes many forms such as foolproof 'jigs' or very tight tolerances on manufacturing components so that defects do not occur. There is a very large opportunity in the process industries to transfer this type of thinking to the existing process designs. Examples already occur without being under the heading of 'Poka Yoke' — for example, building plants without flanges and removing the source of valves so that they cannot leak and building in bursting discs to prevent the release of material. The bursting disc is, however, a poor example since the design approach recognizes that the problem will occur and simply compensates for it, rather than removing it at source. The design is intended to minimize the need for any shutdowns and to even extend the concept of throw-away plants that never require maintenance. This is reflected in the fact many of the incidents that occur in the process industries arise at times of shutdowns and maintenance.

While a long list could be developed, the basic message is that other industries have found means of designing their manufacturing process so that it is impossible to make mistakes. Extending this concept to the process industries offers tremendous scope.

10.2.1.2 Manless plants

The safety, health and environmental risks from process plants affect two populations. Those who operate the plants and those who, for whatever reason, are adjacent to a plant at the time of an incident. The concept of a manless plant is a means of reducing the risk to those who operate the plant. If there are no people present on the plant during operation then the risk of injury and health hazards reduces to an extremely small level, since there is only any risk when people have to visit the plant for maintenance. Companies in the world are already targeting to achieve 'lights out' operation at night. Subsequently, there is no reason why process manufacturing plants require operators. The technology is known, full automation is possible and all the components exist. However, to achieve manless operation will require the full exploitation of existing known technology.

For example, a full dynamic model of the plant that is used in conjunction with model-based predictive control and multivariant statistical process control needs to always ensure that the plant is under control throughout its total operating range. In addition, the techniques used in other industries for the intelligent monitoring of equipment for potential mechanical failures will need to be transferred to the process industries in areas such as non-destructive testing, ultrasonic diagnostics and even acoustic measurements to listen to the state of the plant. Furthermore, the plant is required to be able to move to a safe or shutdown situation automatically in the event of the slightest doubt occurring. This will allow the maintenance team to visit the plant and resolve any issues. While it may sound far-fetched, given that in the present day the prime purpose of the operators is to keep the plant in a safe condition, it is technologically not too far away in moving to a manless operation. The difficulty is one of culture rather than technology.

10.2.1.3 Intensification

The concepts of intensifying process operations have been developed over the last 20 years[17–19]. The basic principles are that by minimizing the volume and reducing the size of all the channels, plus the possible use of high gravitational forces, moves the basic components of mass and heat transfer into regimes of the Reynolds number where extremely high performance is achieved. Through this high performance, the process conversion efficiency is radically improved in a much reduced volume. This reduction in volume has gone all the way from a high 'g' distillation column which may be 100 times smaller than a conventional column, through to the ultimate of manufacturing chemicals on a micro chip, which is now increasingly being recognized and published as a possibility. One of the many features of intensification is that the contained volume of material within the process is radically reduced. Less volume reduces the risk should a loss of containment in any form occur. Ultimately, a process manufacturing unit could consist of 256 manufacturing processes on a 6" disk of silicon. If one of the processes leak, then the resulting emission would be easily contained and the risk of any incident removed. Hence, intensification is one of the inherent strands that bring about the possibility of a zero safety, health and environmental target.

10.2.1.4 Future targets

Given that the routes of delivering zero safety, health and environmental performance have been identified technically, it is our view that these will be achieved within five years, hence the world-class manufacturing targets for safety, health and environmental incidents will move to zero.

10.2.2 Customer service

This is an area where currently the world-class manufacturing metrics are determined by the consumers. It is already the case in some industries dealing direct with consumers that they are demanding defect rates measured in less that 100 parts per 1,000,000, with OTIF deliveries in excess of 99.7% at the time the customer specifies. These types of performance measures will move to the process industries in general. They will arrive either because the customers demand it or because the manufacturers perceive this as a potential competitive advantage. There are already signs that companies such as Allied Signal, General Electrica and ICI are moving to adopt the principles of six sigma. The concept of six sigma was developed in the early 1980s. It states that the defect rates of all products should achieve a six sigma level of performance as defined in statistical terms. This is described in Figure 10.3.

Achieving six sigma relies on the number of defects in a product being less than four defects per 1,000,000 products. A product may be an order, a drum of chemicals, stock-keeping units or some other product. This target is extremely demanding, but already some companies are aiming to achieve it and this is in the area where again process technology can play a significant role. The fishbone diagram in Figure 10.4 illustrates some of the areas of process technology that will contribute to achieving six sigma performance. Each of these areas and their potential is briefly described.

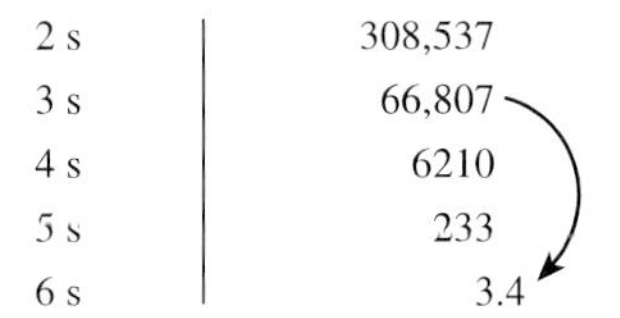

Figure 10.3 What is six sigma? (reproduced by permission of ICI)

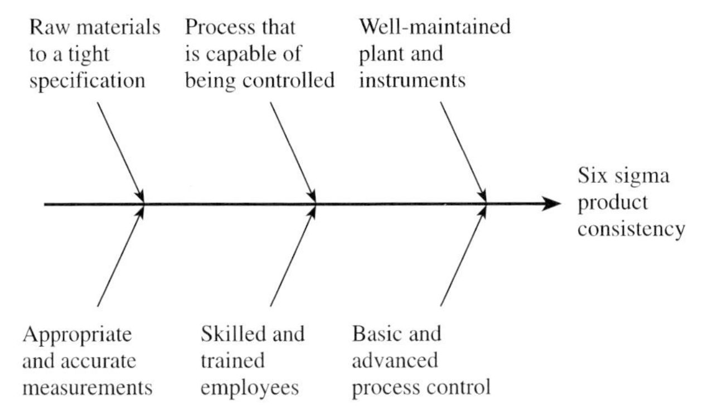

Figure 10.4 Factors which affect overall plant/product consistency (reproduced by permission of ICI)

Delivering this performance through process engineering involves many aspects:

- Design for manufacture:
— build world-class manufacturing metrics into the design process
— statistical performance specifications on all equipment;
- Full process systems modelling and control:
— fully automatic start-up, shutdown and fail-safe
— intelligent measurement
— multivariate statistical process control[20];
- Demanding supplier performance:
— six sigma specifications on all suppliers;
- Intensive training of all people:
— dynamics and SPC in all undergraduate and operator course.

Some of these directions are described in more detail.

10.2.2.1 Supplier performance

This is the least technical area of the contributors. The principle is that all the raw materials, whether chemicals or maintenance parts, are delivered to an extremely high tolerance so that no defects enter the process from the supply side. The process technology challenge here is to define the tight specifications. This will require an intimate knowledge of the cause and effect behaviour of the process. It is an area where failure mode effect analysis (FMEA) is particularly effective. For chemicals, this will demand approximately six sigma-type specifications on the ingredient suppliers, and on the mechanical equipment. It is about having a much tighter specification on the components that are used. This will include very tight tolerances on the material and its control.

This is one of the areas that will drive all the suppliers in the process manufacturing chain, many of whom are other process companies, to dramatically improve their defect rate performance.

10.2.2.2 Training

Again, this may seem a surprising topic to include in this area. The evidence from other industries is that to reduce the defect rates demands a radical improvement in the training of all those involved in designing, maintaining and operating a plant either directly or indirectly and, as has happened in the vehicle industry, a move to computer diagnostics and maintenance procedures[21]. Typical training days for all employees would exceed 12 days per year in leading manufacturers. By building much more knowledge and wisdom into the maintenance and operating procedures, the scope for errors and defects will reduce. The technical opportunity here is in knowledge management to ensure experience is always retained and built into operating processes. It is probable, as plants move towards manless operations, that this diagnostic work will be done both remotely and automatically.

10.2.2.3 Process modelling and control

Given that the quality of the feed stocks is defect free, and the diagnostics are available, this will be the area where the biggest impact will become apparent. It is already possible to dynamically model any process plant be it a large continuous one or a multi-stage batch manufacturing operation. Much of the work at the Imperial College Process Systems Centre cover this area (see Figure 10.5)[22].

It is further possible to optimize these models continuously within the power of existing computers. As the concepts of six sigma develop, process control will move increasingly to statistical process control and hence the area

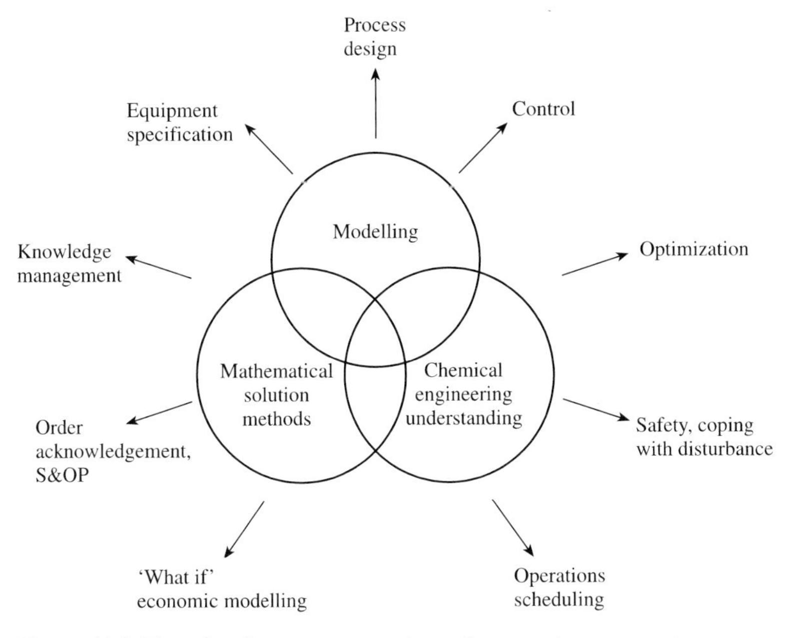

Figure 10.5 The role of process systems in performance improvement

of multivariants statistical process analysis[23–25] will grow in importance, particularly in batch areas where the operating window is small and will move through the batch process. Techniques are already being developed that will allow the operating window to be continuously monitored and measured throughout the batch cycle. The consequence of this will be a much tighter control of the operations and hence lower defects.

10.2.2.4 Customer service

Based on the above suggestions, it is our contention that world-class standards for customer service will move to defect rates that are less than four parts per 1,000,000. Process technology, particularly process control and modelling, will have a major impact in this area.

10.2.3 Plant reliability

This book has introduced the concept of overall equipment effectiveness and suggested that world-class standards are in the order of 85% for a batch type plant and 95% for a continuous plant. The concept of OEE may be new to many of the readers in the process industries, and certainly the targets suggested may at first sight seem unachievable. It is not uncommon for an existing process plant to be operating with an OEE of less than 50%. The question is how much higher OEE could be achieved, how could process technology influence that value, and what will be the long-term target?

It is suggested that an OEE of 99% on the critical equipment will be the future target. This can be achieved by:

- Quality rate > 99.9%:

— six sigma performance;

- Product rate > 99.5%:

— fully automatic start-up, shutdown and fail-safe
— intelligent measurement
— total accurate dynamic models
— multivariate statistical process control;

- Reliability > 99.5%:

— design for success, not failure
— predictive maintenance;

- Batch processes:

— pipeless plants: change the rules.

Some of these are described in more detail.

10.2.3.1 Flexibility versus reliability

The debate which often occurs in the process industry is the appropriateness of OEE to a particular industry. If the real desire of the industry is to have flexibility and to always be able to support the customer, then it may deliberately build into the plant spare capacity to meet this requirement. The consequence at first sight is that the OEE is low while the flexibility is high. This is, however, not a conflict. It is possible to handle this by defining the OEE for the time when the plant is operating or for the rate it is operating at. The ideal situation is that a high OEE is achieved at all the rates, hence when the demand is high the plant has a very high availability. When the rates are low, the plant is shutdown between demands. In this section, OEE is used to describe the performance when the plant is operating at the production rate which the market has determined. There are three components of OEE and each has a role to play.

10.2.3.2 Quality rate

This is the amount of product that is right first time without adjustments, recycles and so on. To achieve the six sigma performance described previously, it is necessary to achieve a very high right-first-time rate. All the steps described will move it in that direction and one would expect the right first time to arise.

There are, however, a number of processes, particularly of a batch-type nature, where the basic reaction kinetics lead to by-products which both reduce the right first time and necessitate separation and purification causing environmental issues. For many reasons, it is highly desirable to remove the by-products of any reaction at the reaction stage rather than later. Achieving this demands a synergy between the number of technologies — for example, intensification which often increases the actual reaction conversion efficiency; sophisticated measurement such as on-line raman spectroscopy to monitor the by-products as formed without adjustments and recycles, and highly-specific catalysis which regulates both the design products and the bioproducts. There are many signs in the literature of increasing specificity among catalysis.

Reactions within material at its critical phase such as carbon dioxide offers the potential for highly-controlled reactions and the radical removal and recycling of by-products into reaction products. Finally, there is scope for moving by-products into main products. For example, many of the chemicals that are used in fragrances in very small amounts appear as by-products in other chemicals which are made in large quantities. Identifying the potential synergy between what could at first sight appear as different parts of the process industries offers a further opportunity to convert a waste product into a useful one to the mutual benefit of all concerned. For all these potential areas of technology, the suggestion is that the right-first-time area should be able to move into the region of 99.57% or more.

10.2.3.3 Product rate

The product rate is defined as the average rate that the plant operates divided by the best that has ever been achieved. In a plant which deliberately has built in headroom to give flexibility, the product rate moves to one which is the average rate divided by the market rate. There is no reason why this should not be 100%. This will be equivalent to designing and operating a plant where the plant rate is set by simply moving a dial on the control panel. The user specifies the market rate and the plant does the rest.

The technology for this exists already in automation and control and, in addition, the tools to optimize the move from one rate to the other also exists. The computing power exists and hence a product rate of 100% is quite achievable in both an intensive plant where the maximum rate is the product rate and a plant with headroom where the product rate is defined by the market rate.

10.2.3.4 Mechanical and process availability

As a reminder, the availability is defined as the hours the plant operates divided by the number of hours in a year. For the reasons outlined in the previous sections, the process availability should, with technology and advances, achieve 100%.

The remaining issue is the mechanical availability. There is currently variation in the performance achieved. Similar plants around the world have periods between maintenance that vary from six months to six years. In other industries, the concept of availability is moved to one where unavailability is both predictable in frequency and never a cause for lost process opportunities. The techniques used to achieve this are very tight tolerances on all raw materials, routine non-invasive maintenance of equipment, intelligent diagnostics and the use of new materials. All these are well-known techniques that are applicable to the process industries. A further feature, however, will be the move to smaller manufacturing plants. With the principle of smaller distributed process manufacturing, the economies of scale will move from one of making large plants to one making many of the same smaller process plants. The learning implicit in repeat engineering, plus the use of smaller equipment which will inherently be under 'less strain', will increase their reliability. The ultimate is the plants on the microchip which will probably have no moving parts reducing the potential for unreliability.

It is from a combination of these techniques, plus the learning that has been achieved in other industries, such as aircraft and cars, which suggests that mechanical availability in excess of 99.8% is possible. This is equivalent to suggesting the plant is shut down for 18 hours per year.

10.2.3.5 Reliability

The consequence of this analysis is to suggest that within five years the world-class target for OEE for a process plant, whether continuous or batch, will rise to approach 99.5%.

10.2.4 Operational excellence

While all the above factors contribute to operational excellence, one measure that is particularly relevant financially is the stock turn. This is defined as the annual turnover at the plant divided by the value of all the stock, raw material, work in progress and finished goods. The arguments used already will dramatically reduce the amount of work in progress for all the reasons of safety, environmental risks, reliability and so on. Many factors affect the stock turn. Stock turn of more than 50 can be achieved by:

- zero SHE targets;
- quality — six sigma;
- OEE > 99%;
- manufacturing velocity > 25%.

As attention focuses on supplier quality, the customer will increasingly dcmand just-in-timc dclivcry of raw matcrials. On many chcmical plants, this is already the case where the raw material is delivered through a pipe. This is the ultimate just in time, since the material is consumed the moment it is delivered and is paid for at the point of delivery. Where this is not available, process manufacturers will seek to make and achieve an equivalent performance by either moving to consignment stock where the supplier holds the material until it is used or to some rotating delivery system such as used in the vehicle part industry. The driver to this is an increasing focus over the next five years on the supply chain as the key business process to improve the competitiveness of the process industries.

The main component of the stock turn is the finished goods stock. Currently, the plants with the minimum finished goods stock are those where the chemical is either very hazardous or where it again is distributed via a pipe immediately to a consumer. The good news about the pipe arrangement is that it achieves a high stock turn, but the bad news is the supplying plant is very vulnerable to any failures in the consuming plant. As the number of stock-keeping units (SKU) that the plant produces increases, it is conventional logic that the amount of stock necessary also increases. The driver for this is the principle of a minimum economic batch size. A consequence is a need to rotate the process equipment through these batches in a manufacturing wheel. If, however, the inherent volume within a process diminishes and the process and mechanical reliability increases, it is possible to conceive of the situation where the minimum economic batch size becomes a single stock-keeping unit and the transition between the stock-keeping unit becomes virtually zero. In these circumstances, even the process industry can conceive a move to a just-in-time delivery system where the order is received and the process manufactures a specific order in the volume specification required by the customer.

This is a consequence of the technology described above plus the pressure on the industry to improve its supply chain performance. The resulting effect is for the finished goods stock level on most process plants to reduce to a level that is measured in terms of one or two day orders. This suggests that the total stock in the chain given the raw materials, finished goods, engineers' spares and work in progress could approach four or five days given a stock turn in excess of 50. This is measured in terms of stock turn which will be in excess of 50 in five years time.

10.2.5 Financial measures

There are many methods which can be used to describe the impact of improving world-class manufacturing measures. For the purposes of this argument, the one used is the added value of employee. This is described as the sales value from the manufacturing plant less the fixed and variable costs and divided by the total number of employees associated with manufacturing and distributing the products. It was indicated in earlier chapters that a world-class figure for a capital-intensive continuous-type plant would currently be around £400K per person and for a batch or discontinuous type process where this issue is on working capital fixed capital, the world-class figure would be nearer to £200K per employee.

The impact of moving, through process engineering, would be an added value per employee of nearer £2,000,000 for continuous plants and £1,000,000 per employee for batch plants.

These show very significant differences from today and illustrate the sensitivity of the businesses' financial performance to its manufacturing performance. As the leaders in the process industries move over the next five years to these world-class manufacturing metrics, through process technology, the gap between the winners and the losers will continue to widen and unfortunately those that do not move will eventually disappear, just as has happened in other industries.

10.3 Future world-class metrics

Based on the above discussion, Table 10.1 summarizes what could be a set of world-class process manufacturing metrics in five years time and compares them with the current world-class metrics and the typical average performance in the process industries.

In planning future investments in new assets, which often takes one to three years to build, the future world-class metrics are suggested as a design target.

10.4 Summary

This chapter gives an idea of how some areas like process engineering/technology, using existing knowledge and future research, could and probably will move the world-class benchmarks for the process industries. Suggestions are made regarding how the various benchmarking measures could be improved. It concludes with a set of manufacturing benchmarks for the process industries for the next five years. We would like to emphasize that these benchmarks are speculative and are provided only as a stimulus to future thought.

Table 10.1 The world-class standards determined by process technology

	Poor	Good	World class today	World class tomorrow
m hrs/reportable injuries	0.01	0.1	> 1	> 10
Environmental incidents				
OTIF%	40%	99.9%	> 99%	> 99.9%
Customer complaints (% of orders)	6%	0.01%	< 0.1%	< 0.0004%
Production rate	60%	99%	> 90%	> 99%
Quality rate	45%	99%	> 99%	100%
Availability %	70%	96%	> 95%	100%
OEE	20%	94%	> 85%	> 99%
Process capability (*Cpk*)	0.6	1.5	> 2	> 6
Stock turn	4	19	> 25	> 50
Manufacturing velocity	0.1%	1%	> 8%	> 25%
Supply chain costs as % of sales	22%	12%	< 9%	< 6%
Absenteeism	10%	0.8%	< 1%	< 1%
Added value per employee	£10K	£200K	£400K	> £2000K

References in Chapter 10

1. Ahmad, M.M., 1996, Next generation process manufacturing systems, *Proc of 6th Int Conf on Flexible Automation and Intelligent Manufacturing* (Begell House Publishers, New York, USA).
2. Benson, R.S., 1997, UKACE lecture: Process control — an exciting future, *Computing and Control*, September.
3. Benson, R.S., 1989, Process systems engineering: past, present and a personal view of the future, *Computers in Chemical Engineering*, 13(11/12).
4. Benson, R.S. and Ponton, J.W., 1993, Process miniaturisation — the route to total environment acceptability, *Trans IChemE, Part A, Chem Eng Res Des*, 71(A2): 160–168.
5. Gardiner, K.M., 1998, Manufacturing processes for the 21st century, preface, *Proceedings of the 12th Conference with Industry, Centre for Manufacturing Systems Engineering, Leigh University, USA, May*.
6. IACOCCA Institute, 1991, *21st Century Manufacturing Enterprise Strategy* (Lehigh University, Bethlehem, USA).
7. IChemE, 1998, *Future Life* (IChemE, UK).

8. IChemE, 1996, *Process Competitiveness in the 21st Century* (IChemE, UK).
9. Naisbitt, J., 1982, *Megatrends — Ten New Directions Transforming our Lives* (Warner Books).
10. Naisbitt, J., 1994, *Global Paradox — the Bigger the World Economy, the More Powerful its Smallest Players* (William Morrow and Company, New York, USA).
11. DTI, 1994, *Competitiveness — Helping Business to Win* (HMSO, UK).
12. European Communities Commission, 1993, *White Paper: Growth, Competitiveness, Employment — The challenges and ways forward into the 21st century.*
13. UK Government, 1994, *White Paper on Competitiveness* (HMSO, UK).
14. Office of Science and Technology, 1995, *Progress through Partnership: Report from the Steering Group of the Technology Foresight Programme* (HMSO, UK, ISBN 0 11 430130 1).
15. Fowler, J., 1995, STEP for data management, exchange and sharing, *Technology Appraisals*.
16. Gardiner, K.M., 1996, An integrated design strategy for future manufacturing systems, *J of Manufacturing Systems*, 15(1): 52–61.
17. Ramshaw, C., 1983, HIGEE distillation — an example of process intensification, *The Chemical Engineer*, February, pp. 13–14.
18. Burns, J.R. and Ramshaw, C., 1996, Process intensification: visual study of liquid maldistribution in rotating packed beds, *Chem Eng Sci*, 51(8):1347–1352.
19. Jachuck, R.J., Lee, J., Kolokotsa, D., Ramshaw, C., Valachis, P. and Yanniotis, S., 1997, Process intensification for energy saving, *Journal of Applied Thermal Engineering*, 17(8–10): 861–867.
20. Gurden, S., Martin, E.B. and Morris, A.J., 1998, Multivariate statistical process control of a industrial process, *Chemometrics and Intelligent Laboratory Systems, (in press)*.
21. Jordan, P., 1998, Steps to improvement — developing people, *Manufacturing Engineer*, 77(5).
22. Mohideen, M.J., Perkins, J.D. and Pistikopoulos, E.N., 1996, Optimal synthesis and design of dynamic systems under uncertainty, *Computers in Chemical Engineering*, 20, Suppl, pp S895–S900, S0098–1354.
23. Wronski, G. and Wilson, R.G., 1984, Modernisation of control systems to maintain plant reliability and economy in a flexible operating regime, *Proceedings of the Institution of Mechanical Engineers, Part A — Journal of Power and Energy*, 198(1): 7–12.
24. Nikolaon, M., 1998, Process measurement and control — industry needs, *Computers and Chemical Engineering*, 23(2): 229–245.
25. Koelsch, R., 1999, Profitable process control, *Modelling Systems*, 57(1): 32–35.

Conclusions

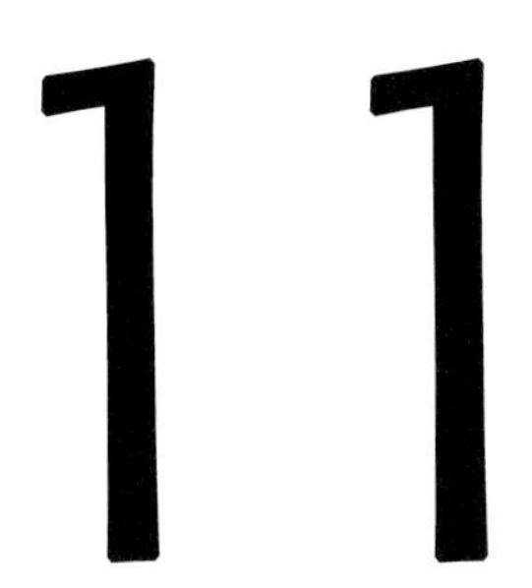

This book has provided a framework for benchmarking process manufacturing. Through this framework it is suggested that the opportunity to improve the performance of the majority of the existing assets is significant. The work presented provides guidelines and methodology on how to measure and benchmark the manufacturing performance of a process plant as a first step on the journey to performance and competitiveness improvement.

- The performance measures for process manufacturing have been identified — that is, OTIF, customer complaints, due date reliability, adherence to production plan, stock turn, OEE, quality rate, availability, people motivation and SHE. These measures demonstrate the practicality of measuring the manufacturing performance of any process plant. The nature and importance of these measures is dependent on the nature of the industry and market.
- The sources of appropriate benchmarking targets for the various measures for world-class performance are identified. It represents what is the best performance or practice anywhere in the world.
- Through benchmarking, gaps are identified and an outline of the procedure to quantify these gaps which exist is provided such as added value per employee, hidden plant, variable cost opportunity, fixed cost per tonne, maintenance cost, potential cost savings opportunity and SHE. All the tools focus on quantifying a gap financially which sets out the priorities and justifies the case for continuous improvement.
- Once the gaps are identified, guidelines are given for companies on how to make process improvements which are realistic and achievable. It is always a difficult decision to make on which tools and technology are most appropriate for performance improvement. Guidelines are provided on this difficult subject of technology management.
- Manufacturing excellence practices are discussed and their relevance to performance improvement established. Several alternative models have been discussed and compared based on their common characteristics which are that they can be scored and plotted. It is also understood that it will never be

possible to improve all the practices all of the time. It is better to focus on one or two in a particular time period.

- Contributions of people in manufacturing performance improvement are critical and winning companies know that people make the difference. Therefore the areas which affect people performance are identified such as change management, empowerment, learning organization, innovation, winning strategy, logistics, appropriate systems to support business and learning organization culture.
- The interpretation and analysis of benchmarking information is presented by using a spider diagram and producing the correlation between practice and performance; this helps to display the priorities. Graphical representation communicates the message on prioritization and is a good method to share information between companies. Case studies are provided for companies where the benchmarking methodology has been used. It is useful information for the readers to know what to expect from a strategy and analysis of the results.
- Further developments such as agile manufacturing and benchmarking are discussed and some thought-provoking ideas were put forward. It is argued that process manufacturing will change the nature of the process industry from one of centralized manufacturer in relatively large plants on a few sites that break chemicals into sub-molecules to a distributed manufacturer that has processes in the manufacturing plants and eventually in the consumer's house. Throughout this chain the scale of manufacturing diminishes, the scope of technology is different and the opportunity increasingly arises in the manufacturing of process equipment.
- A set of world-class process manufacturing metrics that will be achieved within five years is also suggested. These benchmarks set the targets for future design and also indicate the size of the opportunity and the threat that will rise to the process industries in moving to world-class process manufacturing.
- A consequence of the drive to benchmarking on process manufacturing will inevitably be for less people actually employed in the manufacturing elements of the industry. Smaller plants, manless operation, full automation of operation and diagnostics will all lead to less people being directly employed but will dramatically improve the SHE performance. The threat of this in the reduction of manufacturing jobs is offset by the opportunities it creates to design and manufacture the many smaller distributed manufacturing plants. These will be high technology and mass-produced with new materials and minimum waste considering the total life cycle issues from concept to decommissioning.

This is an exciting opportunity for process engineers but it is a very different one which the existing process industries must not fail to grasp.

Appendix 1

Benchmarking requires reporting performance against a common and agreed set of definitions. The purpose of this Appendix is to collate the definitions of the most widely-used manufacturing, supply chain and SHE. It is not exhaustive.

A1.1 Manufacturing measures

Turnover

This is defined as the value of all services rendered and goods or equipment sold in the UK and overseas, inclusive of:

- work-in-progress adjustments including any toll manufacture;
- related companies turnover.

Number of employees

This is defined as the average number of employees during the accounting year including:

- for holding companies, employees from consolidated subsidiaries;
- all temporary or contract employees in terms of full-time equivalent employees;
- any non-plant based employees who are fully or partially allocated to the site.

Table A1.1, page 128, may help the calculation.

Variable costs

The cost of all bought-in materials and services required for the manufacture of the finished product, including all distribution costs.

Maintenance costs

The costs incurred annually in maintaining the assets.

Table A1.1 Calculating the number of employees

	Department	Examples of work areas	Number of full-time employees
A	Plant management and supervisory	(a)	
B	Technical support	Manufacturing Industrial and/or process engineering Support staff (b)	
C	Materials management	Planning, scheduling Order processing Production and inventory control Purchasing Warehouse operations (c)	
D	Other indirect support	(d)	
E	Direct value adding production employees	Component of intermediate manufacturing Assembly and/or Packaging (e)	
F	Total of production-related employees ($A + B + C + D + E = F$)	(f)	
	Design	Product design and development	
	Other	All other support activities such as personnel and accounting	
	Total number of employees		

Fixed costs

The costs of salaries, maintenance, rent, rates and so on that are included within the fixed costs for the site.

Total cost of maintenance

The maintenance costs plus the opportunity costs of lost production.

Fixed assets

Tangible assets whose life is spread over a number of sections including property, plant fixtures and fittings, office equipment and motor vehicles all at their net book value (this will include leased and capitalized assets and, for some industries, assets held for renting or hiring out).

Manufacturing added value per employee

This is calculated by taking the total value of the plant or business sales; subtracting the cost of all purchased raw materials, services and other variable and fixed costs and overheads and dividing the result by the total number of employees (in terms of full-time equivalent employees include all temporary or contract employees). The calculation for added value per employee:

$$\frac{\text{(Annual turnover} - \text{variable costs} - \text{fixed costs and overheads)}}{\text{Number of employees}}$$

OTIF

'On time in full' delivery performance — the percentage of all deliveries that are made to the final customer to a mutually agreed date and location, defect free in all aspects. This includes the product quality, order size, packaging and support paperwork.

Adherence to production plan – %

Adherence to the production plan is checking that there is a sales and operations plan (S&OP) in place and it measures how closely production comes to meeting the plan for the previous planning period (typically one month). Deviations from the plan may be due to equipment failure which will be captured elsewhere in the assessment but may also be due to poor forecasting and indicate opportunities for improving the demand management process. If an S&OP is used but adherence is not measured then use an estimate otherwise record n/a and score 0.

Product rate

A measure of the production performance, with 100% achieved by running the plant at the maximum proven rate (MPR) at all times when the plant is not totally shut down.

Maximum proven rate (MPR)

The best production rate on good material achieved over a representative period (typically one to seven days) with no rate reductions, breakdowns or shutdowns. For a batch plant, the rate is an average based on the best achieved cycle times over a representative period.

No demand

If the production unit is shut down due to a lack of demand in the market this period is recorded separately. The calculation method for OEE would include this period as being available to run at full rate — that is, this time does not reduce the achieved OEE.

Losses measured here are all deviations from the MPR, including rate losses:

- before and after shutdowns;
- quality control issues;
- lack of operating consistency;
- upstream/downstream stock issues.

Note that a total shutdown due to lack of sales would not reduce OEE, but reducing the production rate for the same reason would. It is important that plant production records accurately reflect the reasons for any rate reduction. In principle, rate reductions for stock effects can then be reported separately.

The pure calculation of product rate is:

- for losses measured in time,

$$\frac{(8760 - \text{time lost to s/d}) \times \text{MPR} - \text{tonnage lost due to being below MPR}}{(8760 - \text{time lost to s/d}) \times \text{MPR}}$$

- for losses measured in tonnes,

$$\frac{(8760 \times \text{MPR}) - P \text{ lost to s/d} - P \text{ lost due to being below MPR}}{(8760 \times \text{MPR}) - P \text{ lost to s/d}}$$

where P is production.

It is often the case, however, that the actual losses due to being below the MPR are not accurately recorded. In this case, the losses can be obtained by subtraction, assuming that losses due to shutdown and losses due to quality failures (see next section) are known, which is normally true. Also, for this case, any periods of no demand must be added back in, as 'no demand' does not cause a reduction in OEE. The 'pure' calculation, as it is based on losses from ideal, does not need this correction. This alternative method, as it is based on adding back from actual, needs the correction.

In this case, the calculation of product rate is:

$$\frac{(\text{Good production} + \text{potential make in periods of no demand} + \text{QC})}{(8760 \times \text{MPR}) - P \text{ lost to s/d}}$$

where P is production.

Quality rate

A measure of the production performance, with 100% achieved when all production is at planned specification (any grade) that's right first time. Losses are measured as a percentage of potential production, after subtracting losses due to shutdown and rate reduction.

The pure calculation is:

$$\frac{8760 \times (1 - \text{availability}) \times (1 - \text{product rate}) \times \text{MPR} - \text{QC}}{8760 \times (1 - \text{availability}) \times (1 - \text{product rate}) \times \text{MPR}}$$

where QC is production lost due to failed quality control.
A simpler version is:

$$\text{Quality rate} = \frac{\text{good production}}{\text{good production} + \text{failed QC}}$$

Availability

A measure of the production performance identifying losses due to total stoppage of production, whether scheduled or unscheduled. Scheduled includes major shutdowns and shifts during a seven-day week that are not worked for any reason. 100% is achieved by never having a total shutdown at any time; 365 days per year. MPR is defined above.

There are two methods of calculation, both giving the same result.

- if losses due to shutdown are measured by time (hours):

$$\text{Availability} = \frac{8760 - (\text{number of hours of total shutdown})}{8760}$$

- if losses due to shutdown are measured in tonnes:

$$\text{Availability} = \frac{(8760 \times \text{MPR}) - \text{tonnes lost due to shutdown}}{(8760 \times \text{MPR})}$$

(Availability is expressed as a percentage).

OEE (uptime)

Overall equipment effectiveness is the total measure of plant performance. The three components are:

- availability;
- product rate;
- quality rate.

These are multiplied together to give a single figure, expressed as a percentage, that can be compared to any other manufacturing facility of any type, indicating the extent to which the asset is being used. A world-class performance for a batch plant is internationally seen as a figure above 85%. This is based on:

- availability > 95%;
- product rate > 90%;
- quality rate > 99%.

$$\text{OEE} = \text{availability} \times \text{product rate} \times \text{quality rate}$$

So world class is:

$$\begin{aligned}\text{OEE} &\geq 0.95 \times 0.90 \times 0.99 \\ &\geq 0.85\ (85\%)\end{aligned}$$

A world-class operation for a continuous plant is greater than 95%. This is based on:

- availability > 98%;
- product rate > 98%;
- quality rate > 99%.

$$\text{OEE} = \text{availability} \times \text{product rate} \times \text{quality rate}$$

So world class is:

$$\begin{aligned}\text{OEE} &\geq 0.98 \times 0.98 \times 0.99 \\ &\geq 0.95\ (95\%)\end{aligned}$$

Note that, by working through these calculations, the final OEE is also:

$$\text{OEE} = \frac{\text{good production}}{(8760 \times \text{MPR})}$$

which is the definition of uptime! See Table A1.2, page 134, for a full calculation.

Notes on OEE

OEE has a clear definition based on the 'widget'-type industries, using eight losses to capture all deviations from the maximum potential output. These can be readily mapped onto a process industry context as in Table A1.3, page 135. The losses are most easily illustrated with a diagram as in Table A1.4, page 135.

Note that periods of no demand do not affect OEE. All losses are quantified in tonnes, with the OEE defined as:

$$\text{OEE} = \text{availability} \times \text{product rate} \times \text{quality rate}$$

This may be called uptime at utility in some businesses.

Characteristics of stock

The axes for the 'stockprint' are normalized on a 0–10 scale. The assessor should explore each of the following factors to understand the reasons for stock.

The *products* axis is intended to capture the extent to which stock is influenced by the product range particularly where several products are made in a discontinuous fashion on the same assets, and days of cover must be established for each product.

The *process* axis aims to account for stock which is held to cover for uncertainties in making what is planned when it is planned and also components of stock which are due to cover for shutdown.

The *demand* axis is a function of the uncertainties in the ability to forecast customer requirements, customer service expectations and also any components of stock due to seasonal demand.

Supply chain logistics aims to accommodate stock held as raw material, work-in-progress and in the distribution chain and reflect the complexity of the geographic disposition of suppliers and customers.

The *strategic* axis is intended to capture decisions which business might take hold stock strategically because, for example, it feels there is competitive advantage in it or as insurance against major plant outages.

The *financial* axis is intended to account for the likelihood of stock being held due to advantageous trading conditions, price discounts and so on.

Table A1.2 Calculation table for OEE

• Best sustained tonnes per hour		Tes/hr	*A*
or			
• Best sustainable batch time		Hours	*B*
• Batch size		Tes	*C*
Maximum Proven Rate (MPR) = *A* or (*C*/*B*)			*MPR*
Availability			
• Tonnes lost due to total shutdown		Tes	*D*
or			
• Time lost due to total shutdown		Hours	*E*
• Calculate Tes lost ($E \times$ MPR)		Tes	*F*
$\frac{(8760 \times \text{MPR}) - D \text{ or } F)}{(8760 \times \text{MPR})}$		Fraction	*Availability*
Product rate			
• Tonnes lost due to operating below MPR		Tes	*G*
$\frac{(8760 \times \text{MPR}) - (D \text{ or } F) - G}{(8760 \times \text{MPR}) - (D \text{ or } F)}$		Fraction	*Product rate*
Alternative calculation			
• Total good production		Tes	*H*
• Total production that failed QC		Tes	*J*
• Potential production in periods of no demand		Tes	*K*
$\frac{(H + J + K)}{(8760 \times \text{MPR}) - (D \text{ or } F)}$		Fraction	*Product rate*
Quality rate			
• Total good production		Tes	*H*
• Total production that failed QC		Tes	*J*
$H/(H + J)$		Fraction	*Quality rate*
OEE			
OEE = availability × product rate × quality rate		Fraction	

Table A1.3 'Widget'-type industries compared with process industries

'Widget' industry loss	Process industry loss
Scheduled shutdown	Scheduled downtime
Equipment failure	Unscheduled downtime
Set-up and adjustment	Start-up and shutdown, grade change
Cutting blade change	Scheduled shutdown
Start-up	Start-up
Minor stoppage and idling	Combination of unscheduled shutdown and rate loss
Speed	Rate
Defects and rework	Quality

Table A1.4 Illustration of losses

8760 hours at maximum proven rate (MPR)	
Operating time at MPR (Availability)	Losses due to scheduled and unscheduled downtime
Net operating time at MPR (Product Rate)	Losses due to start-up rate, shutdown rate and operating rate reductions
Valuable operating time at MPR (Quality Rate)	Losses due to quality failure

Measures of velocity

$$\frac{\text{Value added time}}{\text{Elapsed time}}$$

Inventory record accuracy

There is an inventory control process in place which provides accurate warehouse, stock room and work-in-process inventory data. It is measured by:

$$\frac{\text{Number of correct items / quantities / locations * (within quantity location)}}{\text{Number of items / quantities / locations * checked}}$$

* if discrete location identification is not required, delete from equation.

Yield/efficiency

For upstream

$$\text{Efficiency} = \frac{\text{actual production (tonnes)}}{\text{theoretical production (tonnes)}} \times 100$$

where theoretical production is:

$$\text{Weight of reactant} \times \text{purity} \times \frac{\text{molecular weight of product}}{\text{molecular weight of reactant}}$$

(The means of measuring purity can vary between sites).

For downstream

$$\text{Yield} = \frac{\text{weight of saleable product (tonnes)}}{\text{weight of raw materials consumed (tonnes)}} \times 100$$

The weight of saleable product should take into account cutting losses within the factory. The weight of raw materials consumed should take into account any recycled materials (grind, depolymerized).

Throughput

For upstream
IPR is the 'instantaneous peak rate' and is the hourly product rate when the plant operates at budget yield, at full rate at 100% uptime.

To calculate it choose the quarter with the highest product rate in the previous two years. For this quarter, examine the daily rate and average the highest 10 days rate to provide the IPR in tonnes per hour.

For downstream
IPR is the calculated production in tonnes within a 24-hour period based on running the assets in the most favourable grade (as decided by the site).

When this value is affected by mix the site may give a throughput which is corrected for mix effects. Do not exclude planned or unplanned outages or shutdowns for hold-ups.

Manpower productivity

$$\text{Manpower productivity} = \frac{\text{total production}}{\text{manufacturing manpower numbers}}$$

Manufacturing manpower numbers include:

- operators/technicians;
- engineering technicians;
- management team (plant manager, site manager, engineers);
- technical teams (process engineers, i/e engineers);
- supervisors;
- SHE, administrative support, lab/analysis teams, personnel;
- long-term (more than six months) contractors (this should include those involved in toll manufacture).

Where people are associated with intermediate process stages — for example, making ACH or operating SAR — their numbers should be divided between products if the site has more than one product variant.

The total should exclude:

- short duration contractors (less than six months);
- resource whose location on the site is for convenience purposes — for example, R&T and regional support groups;
- commercial groups.

Total production is the weight of finished products — for example, MMA, sheet (not intermediates or effluent disposal).

Fixed cost per tonne

$$\text{Fixed cost per tonne} = \frac{\text{total fixed costs}}{\text{total production}}$$

Fixed costs are all manufacturing costs other than variable as defined below expressed in £s (GBP) and divided by the production volume for the month. The costs should exclude depreciation.

Variable cost per tonne

$$\text{VPC} = \frac{\text{variable costs}}{\text{total production}}$$

Variable costs are those which vary according to production rate and includes raw materials, utilities/services (steam, electricity and so on) and packaging/glass.

Sales and operational planning

$$\text{Sales and operational performance} = \frac{\text{total production}}{\text{forecast}} \times 100\ (\%)$$

Other manufacturing metrics

Uptime =

$$\frac{\text{hours available to run at top rate} \times 100}{\text{total hours}}$$

Utility =

$$\frac{(\text{actual tonnes produced}) + (\text{tonnes lost due to lack of demand}) \times 100}{\text{IPR} \times (\text{days in period})}$$

(sometimes quoted to exclude effect of planned shutdowns)

Overtime =

$$\frac{(\text{hours worked beyond normal hours in roster}) \times 100}{(\text{hours in period in roster})}$$

A1.2 Supply chain definitions

Supply chain definitions taken from the supply chain council

For the purposes of analysis, the supply chain is defined as the flow of goods, information and cash from the supplier's supplier to the customer's customer. For more information see http://www.supply-chain.org

Finished goods stock turn

Last year's cost of goods sold (COGS) divided by the average value of last year's finished goods stock (including stock on plants, warehouses, distribution warehouses and so on valued at cost).

Customer complaints

The number of customer complaints divided by the total number of orders (not line items) over the last year. If this number is not available, the cost of complaints and returns may be available and this can be divided by turnover to give a figure. This value-based calculation may well under-estimate the percentage of complaints based on order numbers.

Customer lead-time

The average lead-time consistently achieved during the last year from customer order signature/authorization to customer receipt.

Supply chain costs as a percentage of turnover

Supply chain costs are made up of the following five categories:

- Order management including new product releases, phase in and maintenance, creating customer orders, order entry and maintenance, contract/program and channel/distributor management, order fulfilment, distribution and warehousing, transportation and customer invoicing/accounting;
- Material acquisition including materials management and planning, supplier quality sourcing, incoming freight and duties, receiving and material storage and incoming inspection;
- Inventory carrying cost including opportunity cost of all inventory (average held over last year), shrinkage, obsolescence, insurance and taxes;
- Supply chain related finance and planning costs including costs associated with paying invoices, stock taking, accounts receivable and demand/supply planning costs;
- Supply chain management information systems associated with product management, sourcing/material acquisition, manufacturing planning, execution and order processing, logistics and channel management.

Turnover is the total product revenue before taxes and net of discounts.

Forecast accuracy

Based on monthly value forecasts for three months in advance using the following calculation:

$$\frac{(\text{Forecast sum for 12 months} - \text{sum of variances from actual for 12 months}) \times 100}{\text{Forecast sum}}$$

$$= (A - B) \times 100 / A$$

New product introduction

Average time from project initiation/customer request until a new product is first released to a customer based on all new products released in last year.

Activity-based costing

Percentage total of all costs which are allocated on an activity basis divided by the total cost incurred in the business.

Cash-to-cash cycle

is the sum of the debtor days, raw material days, WIP days, finished good days minus creditor days where:

$$\text{Debtor days} = \frac{\text{average debtors over last year} \times 365}{\text{net turnover}}$$

$$\text{Raw material days} = \frac{\text{average of raw material stocks} \times 365}{\text{annual raw material receipts}}$$

$$\text{WIP days} = \frac{\text{average of WIP stocks} \times 365}{\text{annual value of movement from WIP to finished product}}$$

$$\text{Finished product stock days} = \frac{\text{average of finished product stocks} \times 365}{\text{COGS}}$$

$$\text{Creditor days} = \frac{\text{average creditors over last year} \times 365}{\text{material receipts}}$$

Process capability

Cpk is defined as the most demanding attribute of the most important or critical raw material used. In many cases, this means the *Cpk* of the major volume or value raw material, but in some cases it may be more appropriate to measure the *Cpk* of a raw material which has the highest effect on the finished product quality:

$$Cpk = \frac{\text{lower of (upper spec} - \text{average)}}{\text{3 sigma}}$$

and

$$\frac{\text{(average} - \text{lower spec)}}{\text{3 sigma}}$$

Raw materials stock turn

Last year's raw materials receipts divided by the average value of last year's raw materials stock (including stock on plants and in warehouses). In this definition, raw materials includes consumables such as packaging.

Supplier lead-time

The value weighted average lead-times during the last year from order signature/authorization to receipt.

Cost to serve index

The total cost to serve (TCS) approach, increasingly employed by logistics managers, focuses on the overall cost of processing and delivering orders, which will vary by product and market, trade channel and between individual customers.

The major elements of supply chain costs can be categorized as the cost of manufacturing the product; the cost of holding stock — that is, the financing costs plus warehousing space for inventories of raw materials, WIP and finished goods, including physical handling in the warehouse as well as transportation and all the related administrative costs such as order processing, invoicing and collection. In addition, the TCS concept embraces marketing and sales costs.

A1.3 Safety, health and environment (SHE)

Safety

Definition of accident categories for use in reporting and analysis

Work-related injury accidents — inclusive categories

- Injury accident (constituting the category 'all injury accidents') — an accident which results in a person having treatment for an injury resulting from an event in the work environment — that is, the employer's premises and other situations where employees are engaged in work-related activities or are present as a condition of employment;
- Classified injury accident — an accident which results in a person having treatment for an injury meeting the criteria for classification. These include reportable accidents;
- Reportable injury accident — an accident which is fatal, major or is an 'over three day' injury accident.

Injury accidents — exclusive categories

- Minor injury accident — an accident which is not a classified injury;
- Classified not reportable injury — an accident other than a fatal accident, major injury accident or 'over three day' injury accident which is not a reportable injury accident;
- 'Over three day' injury accident — an accident, other than a fatal or major injury accident, which results in an employee having an injury which incapacitates them from work of a kind which they might reasonably be expected to do for more than three consecutive days (excluding the day of the accident, but including any days which would not have been working days);
- Fatal injury accident — an accident which results in death.

Definitions of terms used in accident statistics

- Average number of employees — the average number of employees in a work area is the average number of full-time and seconded employees in the period. Where applicable, the average number of part-time and casual employees should be included in proportion to their actual hours worked;
- Total hours worked — the actual hours worked is shown by personnel department records;
- All injury rate — The all injury rate is calculated from the formula:

$$\frac{\text{Number of injury accidents}}{\text{Total hours worked}} \times 10^5$$

- Classified injury rate is calculated from the formula:

$$\frac{\text{Number of classified injury accidents}}{\text{Total hours worked}} \times 10^5$$

- Reportable injury accident rate is calculated from the formula:

$$\frac{\text{Number of reportable injury accidents}}{\text{Total hours worked}} \times 10^5$$

This rate replaces the 'lost time accident rate' previously used.

- Fatal accident rate is calculated from the formula:

$$\frac{\text{Number of fatalities}}{\text{Total hours worked}} \times 10^8$$

- Annual all injury incidence is calculated from the formula:

$$\frac{\text{Number of injury accidents in the year}}{\text{Average number of employees}} \times 10^3$$

- Annual reportable injury accident incidence is calculated from the formula:

$$\frac{\text{Number of reportable injury accidents in the year}}{\text{Average number of employees}} \times 10^3$$

- Annual 'over three day' accident duration rate is the average number of working days lost per 'over day' injury accident, excluding the day of the accident, during a year.

The rate is calculated by dividing the total number of days lost (excluding the day of the accident) during the year as a result of 'over three day' accidents no matter when they occurred by the number of 'over three day' accidents which occurred during the period considered:

$$\frac{\text{Total days lost during year}}{\text{Number of 'over the day' accidents during the year}}$$

Terminology

The following definitions are currently in use in the US:

- Lost work case (LWC) — any injury or occupational illness that results is an employee missing the next scheduled shift of work;
- Total recordable cases (TRC) — any diagnosed occupational illness (regardless of treatment) or any injury requiring treatment by a doctor, or any injury requiring treatment with prescription medicines.

Appendix 2

Data collection form

Scope

This data collection form applied to any manufacturing 'plant'; a 'plant' being defined as a relatively self-contained unit with its own management staff which can be identified either by separate facilities, products or management structure.

Purpose

This data collection for the plant will enable opportunities for beneficial improvements identified, and is the basis for developing your own benchmarking form. It is divided into three sections:

1 — business and site information;
2 — opportunities/benefits;
3 — performance assessment.

In some cases, exact numerical values may not be available to answer the questions. In these circumstances, the facilitator should use best judgement to estimate the value. Every effort should be made to complete all the assessments in full before leaving the site.

The potential value for improvements in manufacturing will need to be estimated.

1 Business and site information

Assessment date	
Collected by:	....
Plant name(s) and title	....
Location and country	....
Telephone number(s)	....

Process plant information

Annual turnover at sales value	....
Annual output, tonnes	....
Average price/tonne	....
Number of manufacturing employees	....
Manufacturing employee salary costs	....
Average manufacturing hour/week	....
	....

Manufacturing contacts at this plant	*Name*	*Position*	*Telephone and fax*
	....	....	....
	....	....	....
	....	....	....
	....	....	....

Main product list	....
	....
	....
Number of 'main products'	....
How many 'product variants' such as packaging, distribution and labelling?	
	....
	....
	....
	....

Process style – for example,batch, continuous, and packing line

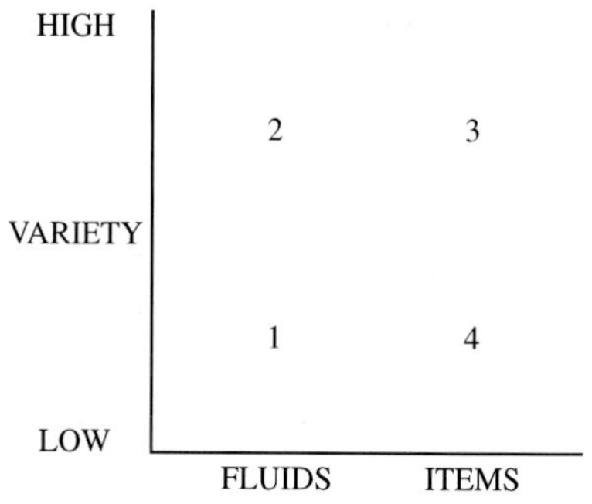

VARIETY	FLUIDS	ITEMS
HIGH	2	3
LOW	1	4

PROCESS STYLE

VARIETY	FLUIDS	ITEMS
HIGH	Recipe Management	Volume and mix Flexibility
LOW	Reliability Optimization	High speed Low cost

DOMINANT DEMANDS

Typically:

1) Single-stream plants processing liquids and different gases. Little product variety.

2) Multi-stream, multi-product plant processing liquids, gases, flowing solids and includes formulation plants.

3) Multi-product plant processing items including small containers of fluids.

4) Single-stream plant producing items with little or no variety.

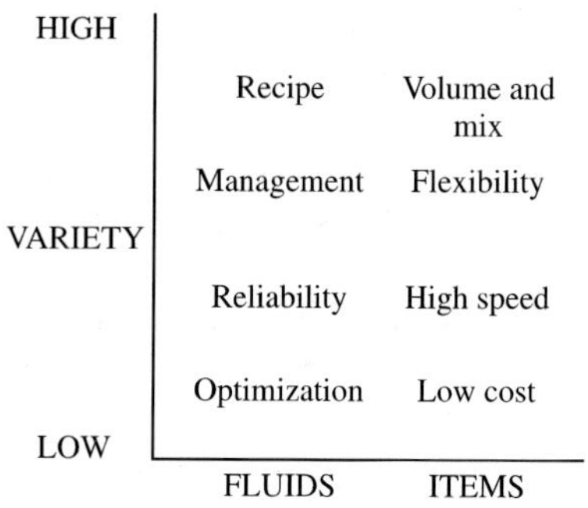

Different process styles place dominant demands on the manufacturing operations.

Those shown illustrate typical characteristic requirements in the four sectors.

Figure A2.1 The main manufacturing transformation stages in the process. Please indicate the process style in each stage.

2 Opportunities/benefits

For each question, please estimate the potential for improvement.

This should be judged against known best practice.

	Circle one	A	B
1. Could the plant output be increased? Comments:	Yes/No		
2. Could the plant capacity be improved by increasing first pass efficiency? Comments:	Yes/No		
3. Could the product consistency be improved? Comments:	Yes/No		
4. Could the plant 'uptime' be improved? Comments:	Yes/No		
5. Are there frequent grade or rate changes, and could average time to reach product specification following such changed be reduced? Comments:	Yes/No		
6. Could the product delivery performance and cost be improved and the final product stocks reduced? Comments:	Yes/No		
7. Could the material and source goods delivery, reliability and quality be improved and the stocks reduced? Comments:	Yes/No		
8. Could manufacturing help the customer to add value? Comments:	Yes/No		
9. Could the total cost of maintenance be reduced? Comments:	Yes/No		
10. Could employee training and empowerment be increased? Comments:	Yes/No		

	Improvement	*Benefit*
Total estimated value of annual improvements, assuming you could sell additional production at current market prices		

A — Estimated potential annual improvement in turnover
B — Estimated potential benefit

Comments:

Plant variable costs per annum	
Plant fixed costs and overheads/annum	
Maintenance costs included in fixed costs	
Plant asset replacement value as is	

These answers will help to focus on the priorities and identify the appropriate tools/techniques/experience and/or training required.

3 Performance assessment

Please use average figures across the plant

Circle relevant number for each criterion.
Enter actual figures in the boxes.
Average

			199	199
Manufacturing added value/employee				
1.	Manufacturing added value per manufacturing employee for the plant Do you routinely measure?	Yes No	£	K ....
Customer service				
2.	% on time in full (OTIF) delivery performance Do you routinely measure?	Yes No	....	....
3.	Adherence to production plan — % Do you measure?	Yes No	....	....
4.	Customers complaints — % of orders delivered. Do you routinely measure?	Yes No	....	....
Reliability and consistency				
5.	Product rate (%) Do you routinely measure?	Yes No	....	....
6.	Quality rate (%) Do you routinely measure?	Yes No	....	....
7.	Do you routinely measure lost time?	Yes No		
	Scheduled downtime — % capacity?	Yes No	....	....
	Unscheduled downtime — % capacity?	Yes No	....	....
	Availability is 100% less the sum of the scheduled and unscheduled downtime		....	....
Calculated OEE				
8.	Maintenance costs as a % of the replacement asset value	Yes No	....	....
Control flexibility				
9.	Process capability Do you routinely measure *Cpk*?	Yes No	....	....
10.	Stocks			
	Value of material and source stocks			
	Value work in progress			
	Value of engineering spares and store items including catalyst			
	Total value of finished goods stocks			
	Total stocks value			

Continued overleaf

Stock turn

Do you routinely measure?	Yes No	....	....
			Average

People

11. Absenteeism (%) Do you routinely measure?	Yes No	....	....

Other important factors

Safety performances:	All injury frequency rates	....	....
Environmental performance		....	....
Average training days/employee		....	....

List of abbreviations

BPR	Business process re-engineering
Cpk	Process specification
EDI	Electronic data interchange
FMEA	Failure mode effect analysis
GMP	Good manufacturing process
JIT	Just in time
MPR	Maximum proven rate
OEE	Overall equipment effectiveness
OTIF	On time in full
Ppk	Product specification
QC	Quality control
RONA	Return on net assets
RT&E	Research, technology and engineering
S&OP	Sales and operational policy
SHE	Safety, health and environment
s/d	Standard shutdown
SMED	Single minute exchange of dies
SPC	Statistical process control
TPM	Total productive maintenance
TQM	Total quality management
WCA	World-class availability
WCE	World-class energy
WCP	World-class product rate
WIP	Work in progress

Index